T0212106

Lecture Notes in Mathematics

C.I.M.E. Foundation Subseries

Volume 2284

More information about this subseries at http://www.springer.com/series/3114

Fondazione C.I.M.E., Firenze

C.I.M.E. stands for *Centro Internazionale Matematico Estivo*, that is, International Mathematical Summer Centre. Conceived in the early fifties, it was born in 1954 in Florence, Italy, and welcomed by the world mathematical community: it continues successfully, year for year, to this day.

Many mathematicians from all over the world have been involved in a way or another in C.I.M.E.'s activities over the years. The main purpose and mode of functioning of the Centre may be summarised as follows: every year, during the summer, sessions on different themes from pure and applied mathematics are offered by application to mathematicians from all countries. A Session is generally based on three or four main courses given by specialists of international renown, plus a certain number of seminars, and is held in an attractive rural location in Italy.

The aim of a C.I.M.E. session is to bring to the attention of younger researchers the origins, development, and perspectives of some very active branch of mathematical research. The topics of the courses are generally of international resonance. The full immersion atmosphere of the courses and the daily exchange among participants are thus an initiation to international collaboration in mathematical research.

C.I.M.E. Director (2002 – 2014)
Pietro Zecca
Dipartimento di Energetica "S. Stecco"
Università di Firenze
Via S. Marta, 3
50139 Florence
Italy
e-mail: zecca@unifi.it

C.I.M.E. Director (2015 –)
Elvira Mascolo
Dipartimento di Matematica "U. Dini"
Università di Firenze
viale G.B. Morgagni 67/A
50134 Florence
Italy
e-mail: mascolo@math.unifi.it

C.I.M.E. Secretary
Paolo Salani
Dipartimento di Matematica "U. Dini"
Università di Firenze
viale G.B. Morgagni 67/A
50134 Florence
Italy
e-mail: salani@math.unifi.it

CIME activity is carried out with the collaboration and financial support
of INdAM (Istituto Nazionale di Alta Matematica)

For more information see CIME's homepage: **http://www.cime.unifi.it**

Guido De Philippis • Xavier Ros-Oton •
Georg S. Weiss

Geometric Measure Theory and Free Boundary Problems

Cetraro, Italy 2019

Matteo Focardi • Emanuele Spadaro
Editors

FONDAZIONE
CIME
ROBERTO CONTI
CENTRO INTERNAZIONALE MATEMATICO ESTIVO
INTERNATIONAL MATHEMATICAL SUMMER CENTER

Authors
Guido De Philippis (iD)
Courant Institute of Mathematical Sciences
New York, NY, USA

Xavier Ros-Oton (iD)
Department of Mathematics
University of Barcelona
Barcelona, Spain

Georg S. Weiss
Faculty of Mathematics
University of Duisburg-Essen
Duisburg, Germany

Editors
Matteo Focardi (iD)
Department of Mathematics and Computer
Science
University of Florence
Florence, Italy

Emanuele Spadaro (iD)
Department of Mathematics
Sapienza University of Rome
Rome, Italy

ISSN 0075-8434   ISSN 1617-9692   (electronic)
Lecture Notes in Mathematics
C.I.M.E. Foundation Subseries
ISBN 978-3-030-65798-7   ISBN 978-3-030-65799-4   (eBook)
https://doi.org/10.1007/978-3-030-65799-4

Mathematics Subject Classification: Primary: 49Q20; Secondary: 35R35

This Springer imprint is published by the registered company Springer Nature Switzerland AG.
The registered company address is: Gewerbestrasse 11, 6330 Cham, Switzerland

Contents

1 **Introduction** 1
  Matteo Focardi and Emanuele Spadaro

2 **A-Free Measures and Applications** 5
  Guido De Philippis

3 **Regularity of Free Boundaries in Obstacle Problems**.................... 37
  Xavier Ros-Oton

4 **Bernoulli Type Free Boundary Problems and Water Waves** 89
  Georg S. Weiss

Chapter 1
Introduction

Matteo Focardi and Emanuele Spadaro

Geometric Measure Theory (GMT) is by now a classical subject, having a well-recognized independent status, but also sharing plentiful natural connections to many other areas of mathematics, such as partial differential equations, the calculus of variations, geometric analysis, and free boundary and phase transition problems, to name just a few. Indeed, apart from the traditional problems arising within the context of GMT (e.g., the multidimensional Plateau problem), the analytic and geometric techniques studied by means of GMT were fundamental in the solutions of many questions in different fields: one of the most striking examples is De Giorgi's solution of Hilbert's 19th problem, which exploited his own ideas on the isoperimetric inequality.

Very recently, many new developments in GMT have emerged and several challenging connections to other problems have been pointed out, such as the solution to Yau's conjecture on minimal surfaces, estimating the measure of the singular set of mass minimizing hypersurfaces and minimizing harmonic maps, the regularity of higher co-dimension mass minimizing currents, estimating the measure of the free boundary for thin obstacle problems, the evolution of grain boundaries and other curvature flows, etc. . .

These and other recent achievements are being further extended and exploited in various directions, and are expected to lead to exciting developments in the future. The CIME summer school in GMT and Applications aimed to join different communities working on deeply related subjects and to introduce PhD students and

M. Focardi (✉)
Department of Mathematics and Computer Science, University of Florence, Florence, Italy
e-mail: matteo.focardi@unifi.it

E. Spadaro
Department of Mathematics, Sapienza University of Rome, Rome, Italy
e-mail: spadaro@mat.uniroma1.it

1

young researchers to the new emerging trends in GMT. The school took place from September 1st to 6th 2019 in Cetraro and consisted of four courses, each course lasting six hours. The high number of participants attending the lectures, about one hundred, witnesses the success of this CIME summer school, which has been partially supported by the ERC-STG Grant n. 59229 HiCoS "Higher Co-dimension Singularities: Minimal Surfaces and the Thin Obstacle Problem".

In the following, we briefly describe the contents of each course.

1.1   A-Free Measures and Applications

Prof. G. De Philippis, New York University, USA

A differential constraint imposes several restricions on the structure of a function or of a measure. Some applications of this instance are presented, among them: compensated compactness, necessary and sufficient conditions for lower semicontinuity of functionals along A-free sequences (A being a first order differential operator with constant coefficients) and some structural theorems for BV and BD functions.

1.2   Existence of Min-Max Minimal Hypersurfaces

Prof. A. Neves, University of Chicago, USA

In the 1970s Gromov introduced the notion of a topological spectrum for a very general class of operators. In particular, it applies to the area functional. In this setting the k-th eigenvalue is denoted by the k-width and eigenfunctions correspond to minimal surfaces whose area is the k-width. The subject was dormant for many years until some years ago Larry Guth recovered Gromov's work and proved some growth estimates for the k-width and Codá-Marques and Neves used it to show that manifolds with positive Ricci curvature had distinct minimal hypersurfaces.

These results have attracted many young researchers, and a promising connection with the Allen Cahn equation has been found. In these lectures the interplay between classical ideas in spectral geometry and GMT has been illustrated, giving an introduction to the recent results on Yau's conjecture about the existence of infinitely many minimal surfaces.

The lecture notes of the course by Prof. Neves are not contained in this volume because they are included in the contribution

- Marques F.C., Neves A. (2020) Applications of MinMax Methods to Geometry. In: Gursky M., Malchiodi A. (eds) Geometric Analysis. Lecture Notes in Mathematics, vol 2263. Springer, Cham. https://doi.org/10.1007/978-3-030-53725-8_2.

The courses taught by Prof. Neves in the two CIME summer schools, covering different but sequential material, are summarized in a single set of lecture notes.

1.3 Regularity of Free Boundaries in Obstacle Problems

Prof. X. Ros-Oton, University of Zürich, Switzerland

The course is concerned with the review of the classical obstacle problem pioneered by the work of Caffarelli, going through the classical results on the existence, the regularity of the solutions and the regularity of the free boundary. In the last part of the course X. Ros-Oton gives an introduction to his recent results on the genericity of regular solutions.

1.4 Bernoulli Type Free Boundary Problems and Water Waves

Prof. G. Weiss, University of Duisburg–Essen, Germany

Bernoulli type free boundary problems play an important role in the free surface flow of water waves. After presenting the basic regularity theory, the course focuses on the analysis of stagnation points for 2d steady water waves subject to gravity in the absence of several hypotheses that had previously been assumed in the literature.

Chapter 2
A-Free Measures and Applications

Guido De Philippis

Abstract These notes record the lectures for the CIME Summer Course held by the author in Cetraro during the week of September 2–6, 2019. The goal is to give an overview of some recent results concerning the structure of PDE-constrained measures.

2.1 Introduction

The main goal of these lectures is to provide a review on old and new results concerning the interplay between oscillation and concentration developed by a sequence of functions (or measures) and a linear PDE constraint. Before providing some motivation for this argument let us introduce a few notation. We will mostly consider a linear homogeneous operator with constant coefficients

$$\mathcal{A} = \sum_{|\alpha|=k} A_\alpha \partial^\alpha \tag{2.1}$$

where

(a) $\alpha = (\alpha_1, \ldots, \alpha_d) \in \mathbb{N}^d$ is a multi-index, $|\alpha| = \alpha_1 + \cdots + \alpha_d$ its length, and $\partial^\alpha = \partial_1^{\alpha_1} \cdots \partial_d^{\alpha_d}$.
(b) $A_\alpha \in \mathrm{Lin}(V, W)$ where V and W are *finite* dimensional linear vector spaces.

The operator acts on V-valued functions and gives back W-valued functions:

$$C_c^\infty(\Omega; V) \ni \varphi \mapsto \mathcal{A}\varphi \in C_c^\infty(\Omega; W) \qquad \Omega \subset \mathbb{R}^d \text{ open}$$

G. De Philippis (✉)
Courant Institute of Mathematical Sciences, New York University, New York, NY, USA
e-mail: guido@cims.nyu.edu

© The Author(s), under exclusive license to Springer Nature Switzerland AG 2021
M. Focardi, E. Spadaro (eds.), *Geometric Measure Theory and Free Boundary Problems*, Lecture Notes in Mathematics 2284,
https://doi.org/10.1007/978-3-030-65799-4_2

and, by duality, it can be extended to distributions

$$\langle \mathcal{A}\mu, \varphi \rangle = \langle \mu, \mathcal{A}^* \varphi \rangle$$

where $\mathcal{A}^* : C_c^\infty(\Omega; W^*) \to C_c^\infty(\Omega; V^*)$ is the adjoint operator and $\langle \cdot, \cdot \rangle$ the duality pairing. We will be mostly interested in the case of *Radon measures*, i.e. order zero distributions, and we will denote the space of V valued Radon measure defined on Ω as $\mathcal{M}(\Omega; V)$, see Appendix A for more details on Radon measures.

We will say that a measure (function) $\mu \in \mathcal{M}(\Omega; V)$ ($u \in L^1(\Omega; V)$) is *$\mathcal{A}$-free* if

$$\mathcal{A}\mu = 0 \qquad (\mathcal{A}u = 0)$$

in the sense of distributions.

2.1.1   Example of Differential Operators

Let us list a few examples of differential operators which will be used in the sequel, the reader is invited to write explicitly the linear operators A_α as in (2.1). In what follows we will denote by $\mathbb{R}^m \otimes \mathbb{R}^n$ the space of $m \times n$ matrixes and by $\mathbb{R}^n \odot \mathbb{R}^n$ the space of $n \times n$ symmetric matrixes.

–   The Laplacian, $\Delta : \mathcal{D}'(\Omega; \mathbb{R}^m) \to \mathcal{D}'(\Omega; \mathbb{R}^m)$

$$(\Delta u)^j = \sum_i \partial_{ii} u^j$$

–   The (row wise) curl operator, curl $: \mathcal{D}'(\Omega; \mathbb{R}^\ell \otimes \mathbb{R}^d) \to \mathcal{D}'\left(\Omega; \mathbb{R}^\ell \otimes \mathbb{R}^{\frac{d(d-1)}{2}}\right)$.

$$(\mathrm{curl}\, u)_{ij}^k = \partial_i u_j^k - \partial_j u_i^k.$$

–   The curl operator, curlcurl $: \mathcal{D}'(\Omega; \mathbb{R}^d \odot \mathbb{R}^d) \to \mathcal{D}'\left(\Omega; \mathbb{R}^d \odot \mathbb{R}^d\right)$

$$(\mathrm{curlcurl}\, u)_{jh} = \sum_i \partial_{ih} u_{ij} + \partial_{ij} u_{ih} - \partial_{jh} u_{ii} - \partial_{ii} u_{jh}.$$

–   The (row wise) divergence, div $: \mathcal{D}'(\Omega; \mathbb{R}^d \otimes \mathbb{R}^d) \to \mathcal{D}'(\Omega; \mathbb{R}^d)$

$$(\mathrm{div}\, u)_j = \sum_i \partial_i u_j^i$$

2.1.2   Motivation

$\mathcal{A}$-free measures naturally appear in several contexts that we briefly describe below:

Differential Complexes

Several differential operators $\mathcal{B}$ are characterized by the existence of another operator $\mathcal{A}$ such that $\ker \mathcal{A} = \operatorname{Im} \mathcal{B}$. For instance

Gradients $v = \nabla u \in \mathbb{R}^\ell \otimes \mathbb{R}^d$ if and only if $\operatorname{curl} v = 0$ (row-wise)

Symmetrized gradients $v = \nabla u + (\nabla u)^T \in (\mathbb{R}^d \otimes \mathbb{R}^d)_{\text{sym}}$ if and only curlcurl $v = 0$. see [14, Example 3.10 (e)].

Exterior differential $v = du$ if and only if $dv = 0$.

Boundaries and cycles   An ℓ-dimensional current T in $\mathbb{R}^n$ is a the boundary of an $\ell + 1$ dimensional current, $T = \partial S$, if and only if $\partial T = 0$, see [12] for an account on the theory of currents.

In this way studying for instance the space of functions of bounded variation, i.e.

$$BV(\mathbb{R}^d; \mathbb{R}^\ell) = \left\{ u : \mathbb{R}^d \to \mathbb{R}^\ell : Du \in \mathcal{M}(\Omega; \mathbb{R}^\ell \otimes \mathbb{R}^d) \right\},$$

is basically equivalent to studying the space of curl-*free measures*

$$\left\{ \mu \in \mathcal{M}(\Omega; \mathbb{R}^\ell \otimes \mathbb{R}^d) : \operatorname{curl} \mu = 0 \right\}.$$

In the same way the study of functions of bounded deformations

$$BD(\mathbb{R}^d) = \left\{ u : \mathbb{R}^d \to \mathbb{R}^d : Eu := Du + Du^T \in \mathcal{M}(\Omega; \mathbb{R}^d \odot \mathbb{R}^d) \right\}$$

boils down to the study of the space of curl curl-*free measures*

$$\left\{ \mu \in \mathcal{M}(\Omega; \mathbb{R}^d \odot \mathbb{R}^d) : \operatorname{curl} \operatorname{curl} \mu = 0 \right\}.$$

Note also that, in all the above statements, the sufficient condition might depend on the topology of the domain while the necessary one is always satisfied.

Non-linear PDE as Linear PDE Coupled with a Pointwise Constraint

This point of view was introduced by Murat and Tartar in the 1970s, [21, 22, 26]. Consider the equation

$$\operatorname{div}(A(\nabla u)) = 0.$$

This can be rephrased as

$$\begin{cases} \operatorname{div} w = 0 \\ \operatorname{curl} z = 0 \\ w = A(z) \end{cases}$$

In the same way a conservation law of the form

$$\partial_t u + \operatorname{div}_x f(u) = 0$$

can be phrased as

$$\partial_t u + \operatorname{div}_x w = 0 \qquad w = f(u).$$

This point of view turned out to be extremely fruitful, in particular it leads to existence results (via the so called *div-curl* lemma) and to *convex integration theory*, a powerful tool to construct pathological examples in PDE.

Inner Variation and the Stress Energy Tensor

When dealing with a variational problem of the form

$$\min \int_\Omega F(\nabla v),$$

Euler Lagrange equations are computed by considering variations of the type $v_\varepsilon = u + \varepsilon \varphi$ around a minimizer u and imposing that

$$\frac{d}{d\varepsilon}\bigg|_{\varepsilon=0} \int_\Omega F(\nabla v_\varepsilon) = 0.$$

By standard computations, this leads to

$$\operatorname{div}(\nabla F(\nabla u)) = 0.$$

One can also perform "inner variations" meaning that one considers variations of the form $v_\varepsilon = u \circ \varphi_\varepsilon^{-1}$ where $\varphi_\varepsilon = x + \varepsilon X$, $X \in C_c^1(\Omega; \mathbb{R}^d)$. In this case:

$$\int_\Omega F(\nabla v_\varepsilon) = \int_\Omega F((\nabla \varphi_\varepsilon^{-1})^T \nabla u \circ \varphi_\varepsilon^{-1})$$

$$= \int_\Omega F((\nabla \varphi_\varepsilon^{-1} \circ \varphi_\varepsilon)^T \nabla u) \det \nabla \varphi_\varepsilon$$

$$= \int_\Omega F((\nabla \varphi_\varepsilon)^{-T} \nabla u) \det \nabla \varphi_\varepsilon.$$

where we have used that $\nabla \varphi_\varepsilon^{-1} \circ \varphi_\varepsilon = (\nabla \varphi_\varepsilon)^{-1}$. By expanding

$$(\nabla \varphi_\varepsilon)^{-T} = \mathrm{Id} - \varepsilon (\nabla X)^T + O(\varepsilon^2) \qquad \det \nabla \varphi_\varepsilon = 1 + \varepsilon \operatorname{div} X + O(\varepsilon^2),$$

one gets

$$\int_\Omega F(\nabla v_\varepsilon) = \int_\Omega F(\nabla u) + \varepsilon \int_\Omega F(\nabla u) \operatorname{div} X - \langle \nabla F(\nabla u), (\nabla X)^T \nabla u \rangle + O(\varepsilon^2)$$

which can be phrased as the fact that the *stress energy tensor*

$$T_u = F(\nabla u) \operatorname{Id} - \nabla u \otimes \nabla F(\nabla u)$$

is (row-wise) divergence free. Note that

$$|T_u| \lesssim |F(\nabla u)| + |\nabla F(\nabla u)||\nabla u|$$

which, for most energies is comparable with $|F(\nabla u)|$. In particular if we assume to have a sequence of minimizers $\{u_j\}$ with uniformly bounded energy:

$$\sup_j \int_\Omega F(\nabla u_j) < +\infty$$

then, by Banach Alaoglu theorem, the corresponding sequence of stress energy tensor will converge weak-$*$ as measures to $T \in \mathcal{M}(\Omega; \mathbb{R}^d \otimes \mathbb{R}^d)$ with $\operatorname{div} T = 0$.

Stationary Varifolds

Similarly we say that a smooth ℓ-dimensional manifold $M \subset \mathbb{R}^d$ is minimal if for all $X \in C_c^1(\mathbb{R}^d; \mathbb{R}^d)$

$$\frac{d}{d\varepsilon}\Big|_{\varepsilon=0} \mathcal{H}^\ell(\varphi_\varepsilon(M)) = 0 \qquad \varphi_\varepsilon(x) = x + \varepsilon X(x).$$

Furthermore:

$$\frac{d}{d\varepsilon}\Big|_{\varepsilon=0} \mathcal{H}^\ell(\varphi_\varepsilon(M)) = \int_M \operatorname{tr}\left(P_{T_x M} \nabla X\right)$$

where $P_{T_x M}$ is the orthogonal projection on $T_x M$. Hence, for a minimal surface M, the measure

$$P_{T_x M} d\mathcal{H}^\ell \llcorner M \in \mathcal{M}(\Omega; (\mathbb{R}^d \otimes \mathbb{R}^d)_{\text{sym}}).$$

is divergence free. If $\{M_j\}$ is a sequence of minimal surfaces with equi-bounded area

$$\sup_j \mathcal{H}^\ell(M_j) < +\infty,$$

by Banach Alaoglu Theorem, $\mathcal{H}^\ell \llcorner M_j \overset{*}{\rightharpoonup} \nu \in \mathcal{M}(\mathbb{R}^d; \mathbb{R}_+)$. Being ν a limit of volume measures restricted on surfaces, it would be reasonable to assume that ν should be "ℓ-dimensional". By standard measure theoretic arguments, [2, Theorem 2.28], on can prove the existence of a positive measure σ and of a function $P \in L^\infty(\sigma)$ such that

$$\nu = P\sigma \qquad \text{and} \quad P(x) = \int P d\tau_x(P) \qquad (2.2)$$

where τ_x is a measure on the space of orthogonal projection onto ℓ-planes. Hence one is reduced to study measures ν of the form (2.2) which satisfy the additional constraint

$$\operatorname{div} \nu = 0. \qquad (2.3)$$

Couples (P, σ) as above are called ℓ dimensional stationary varifolds, see [25] (where a different but equivalent definition is given).

2.2   Oscillation

In this section we give a brief of overview on some known results concerning the interplay between differential inclusions and oscillations. This is a well studied problem with several deep results, we are not going to discuss the topic in full details and we will just present some results and ideas which will be propedeutical for the results in the next section. We refer the reader to [17, 20] for a more detailed account on the theory.

Let us consider the following problem: Let $\{u_j\}$ be a sequence of $\mathcal{A}$-free functions

$$\mathcal{A}u_j = 0$$

weakly converging to u in L^p

$$u_j \rightharpoonup u \qquad \text{in } L^p, \ p \in [1, \infty].$$

When can we say that u_j converges *strongly* to u in L^p?

Note that, if we assume that all the functions have equi-bounded support (as we will do in the following), thanks to Vitali's theorem a weakly converging sequence is strongly converging if and only if the following two conditions are satisfied (the first being in force only for $p \neq \infty$)

– *Non-concentration*:

$$\lim_{|E| \to 0} \sup_j \int_E |u_j|^p = 0$$

– *Non-oscillation*: u_j converges to u in measure:

$$\lim_{j \to \infty} |\{|u_j - u| > \varepsilon\}| \to 0 \qquad \text{for all } \varepsilon > 0.$$

Hence the question boils down to understand how a differential constraint restricts possible oscillation and concentration of a sequence of functions.

There are two natural problems linked with the above question. The first one concerns the behavior of (say) $W^{1,2}$ equi-bounded solutions of non-linear problem like

$$\text{div } A(Du_j) = 0$$

where A is a suitable (non-linear) operator such that $|A(\xi)| \approx |\xi|$. Indeed, while in general, non-linear functional are not continuous with respect to weak convergence, one can prove that

$$u_j \overset{W^{1,2}}{\rightharpoonup} u \qquad \Rightarrow \qquad \text{div } A(Du) = 0.$$

This is a consequence of the *compensated compactness theory* of Murat and Tartar and it is at the basis of several results in PDE, see [11, 26] and references therein. We are not going to prove this result here but we want to stress that the ideas and the results described in the following deeply acknowledge the ideas introduced there.

The second problem concerns the behavior of minimizing sequences in the calculus of variations. Let us consider indeed a variational problem of the form

$$\min_{u \in W_0^{1,p}(\Omega; \mathbb{R}^\ell)} \int_\Omega F(Du) \tag{2.4}$$

where $F : \mathbb{R}^\ell \otimes \mathbb{R}^d \to \mathbb{R}$ is a suitable function with p-growth: $F(\eta) \lesssim (1 + |\eta|^p)$. A natural candidate for a minimizer is a function u such that

$$Du \in K := \{F = \inf F\} \qquad \text{a.e..}$$

so a first question consists in classifying those functions which satisfy the above differential inclusion. Note that according to the discussion done in the previous section, this is equivalent to

$$\begin{cases} \operatorname{curl} v = 0 \\ v \in K. \end{cases} \tag{2.5}$$

Note that it could happen that the boundary condition $u = 0$ on $\partial\Omega$ is incompatible with the inclusion $v \in K$. It could however be that there are sequences $\{u_j\} \subset W_0^{1,p}(\Omega; \mathbb{R}^\ell)$ such that

$$\int_\Omega \operatorname{dist}^p(Du_j, K) \to 0.$$

This clearly implies that there is no solution to (2.4). In this case it becomes natural to understand what is the behavior of minimizing sequences. If one assumes that

$$(F(\eta) - \inf F) \approx \operatorname{dist}^p(\eta, K),$$

these sequences are *approximate solutions* of the differential inclusion (2.5), namely they satisfy

$$\begin{cases} \operatorname{curl} v_j = 0 \\ \operatorname{dist}(v_j, K) \to 0 \quad \text{in } L^p. \end{cases} \tag{2.6}$$

Example 2.1 Let us consider the following minimization problem:

$$\inf_{u \in W_0^{1,4}(Q;\mathbb{R})} \int_Q F(Du),$$

where

$$F(\eta) = \left((\eta_1)^2 - 1\right)^2 + (\eta_2)^2, \qquad \eta = (\eta_1, \eta_2) \in \mathbb{R}^2.$$

and $Q \subset \mathbb{R}^2$ is the unit cube. In this case $K = \{(1,0), (-1,0)\}$ and hence all solutions of (2.5) are of the form

$$v(x_1, x_2) = (\mathbf{1}_E(x_1) - \mathbf{1}_{E^c}(x_1))e_1$$

for some $E \subset [0, 1]$. Since this is incompatible with the 0 boundary condition we infer

$$\int_Q F(Du) > 0 \qquad \text{for all } u \in W_0^{1,4}.$$

Let now f be the 1-periodic extension of $(1/2 - |x - 1/2|)_+$ and let

$$u_j(x_1, x_2) = 2^{-j} \varphi_j(x_2) f(2^j x_1)$$

where $\varphi_j \in C_c^1([0, 1], \mathbb{R})$ and $\varphi_j = 1$ on $[2^{-j}, 1 - 2^{-j}]$. A simple computation gives

$$\int_Q F(Du_j) \leq C(2^{-j} + |\{\varphi_j \neq 1\}|).$$

Hence, by suitably choosing the sequence (φ_j), we obtain that Du_j satisfies (2.6) and in particular

$$\inf_{u \in W_0^{1,4}(Q;\mathbb{R})} \int_Q F(Du) = 0.$$

Note that in the above example all exact solutions of the differential inclusion are 1-dimensional and the approximate solution we have found is more and more 1 dimensional. A natural question is whether this is the behavior of *all* minimizing sequences or if other constructions are possible. To understand this point we propose the following exercise to the reader:

Exercise 2.2 Let $F(\eta) = \text{dist}^2(\eta, \{\text{Id}, -\text{Id}\})$ where $\eta \in \mathbb{R}^2 \otimes \mathbb{R}^2$ and Id is the identity matrix. Prove that

$$\inf_{u \in W_0^{1,2}(Q,\mathbb{R}^2)} \int_Q F(Du) > 0.$$

Let us go back to our initial question and consider two "trivial" instances:

- $\mathcal{A} = \Delta$, in this case the classical Weyl's lemma, see Lemma 4.4, ensures that $u_j \to u$ smoothly and thus also in L^p
- $\mathcal{A} \equiv 0$, in this case no constraint is imposed on the sequence.

The above examples suggest that *ellipticity* of $\mathcal{A}$ should play a role, according to the following

Definition 2.3 Let $\mathcal{A} = \sum_{|\alpha|=k} A_\alpha \partial^\alpha$ be a differential operator. We say that $\mathcal{A}$ is elliptic if its symbol

$$\mathbb{A}(\xi) = \sum_{|\alpha|=k} \xi^\alpha A_\alpha \in \text{Lin}(V, W)$$

is injective for all $\xi \neq 0$. Here $\xi^\alpha = \xi_1^{\alpha_1} \dots \xi_d^{\alpha_d}$.

Let us assume that $\mathcal{A}$ is not elliptic. This implies the existence of $\xi_0 \in \mathbb{R}^d \setminus \{0\}$ and $\lambda \in V \setminus \{0\}$ such that

$$\mathbb{A}(\xi_0)\lambda = 0.$$

It is now an easy computation to verify that the function $u(x) = \lambda h(x \cdot \xi_0)$ is $\mathcal{A}$-free for all $h : \mathbb{R} \to \mathbb{R}$. In this case by taking a sequence h_j which is weakly but not strongly converging one easily obtain a sequence of $\mathcal{A}$-free functions which is weakly but not strongly converging.

In view of this consideration it is natural to give the following definition, first introduced by Murat and Tartar:

Definition 2.4 (Wave Cone) Let $\mathcal{A}$ be an homogeneous differential operator, we define its *wave cone* as

$$\Lambda_{\mathcal{A}} = \bigcup_{|\xi|=1} \operatorname{Ker} \mathbb{A}(\xi).$$

Note that the wave cone is a subset of the *state space* V and it contains those vectors in the image along which the operator fails to be elliptic. One can indeed immediately prove the following characterization

Lemma 2.5 *A vector λ does not belong to the wave cone $\lambda \notin \Lambda_{\mathcal{A}}$ if and only if the operator*

$$\mathcal{B}_\lambda : \mathcal{D}'(\Omega; \mathbb{R}) \to \mathcal{D}'(\Omega; W) \tag{2.7}$$

defined as

$$\mathcal{B}_\lambda(u) = \mathcal{A}(\lambda u) \qquad \text{for all } u \in \mathcal{D}'(\Omega; \mathbb{R}) \tag{2.8}$$

is elliptic according to Definition 2.3.

For future reference we report here the wave cones of some of the operators introduced in the previous section

(i) $\mathcal{A} = \Delta$

$$\operatorname{Ker} \mathbb{A}(\xi) = \{0\} \quad \text{for all } \xi \neq 0 \qquad \Lambda_{\mathcal{A}} = \{0\}.$$

(ii) $\mathcal{A} = \operatorname{curl}$

$$\operatorname{Ker} \mathbb{A}(\xi) = \{a \otimes \xi\} \quad \text{for all } \xi \neq 0 \qquad \Lambda_{\mathcal{A}} = \{a \otimes b \quad a \in \mathbb{R}^\ell, b \in \mathbb{R}^d\}.$$

(iii) $\mathcal{A} = \operatorname{curlcurl}$

$$\operatorname{Ker} \mathbb{A}(\xi) = \{a \odot \xi\} \quad \text{for all } \xi \neq 0 \qquad \Lambda_{\mathcal{A}} = \{a \odot b \quad a, b \in \mathbb{R}^d\}.$$

where $a \odot b = a \otimes b + b \otimes a$.

(iv) $\mathcal{A} = \mathrm{div}$

$$\mathrm{Ker}\,\mathbb{A}(\xi) = \{M \in \mathbb{R}^d \otimes \mathbb{R}^d : M\xi = 0\} \qquad \Lambda_{\mathcal{A}} = \{M \in \mathbb{R}^d \otimes \mathbb{R}^d : \det M = 0\}.$$

The above discussion suggests that the oscillation of a sequence of functions can happen only in the directions of the wave cone. Unfortunately the situation might be more complicated than this and several problem are not yet completely understood, see [17, 20]. However this intuition is correct when the functions take value in a set with two elements (at least asymptotically) and this allows to completely solve the natural generalization of (2.5) and (2.6) when the set K has two elements. Namely we have the following theorem, see [8] and also [5, Theorem 2].

Theorem 2.6 *Let $\Omega \subset \mathbb{R}^d$ be an open, bounded, and connected set, and let $\mathcal{A}$ be as in (2.1). Suppose that $\lambda, \mu \in \mathbb{R}^\ell$ with $\lambda - \mu \notin \Lambda_{\mathcal{A}}$. Then:*

(A) *If $u \in L^\infty(\Omega; \mathbb{R}^\ell)$ is $\mathcal{A}$-free and*

$$u(x) \in \{\lambda, \mu\} \quad \textit{for a.e. } x \in \Omega,$$

then either $u \equiv \lambda$ or $u \equiv \mu$.
(B) *Let $p \in (1, \infty)$ and let $(u_j) \subset L^p(\Omega; \mathbb{R}^\ell)$ be a uniformly norm-bounded sequence of $\mathcal{A}$-free functions. Assume that*

$$\lim_{j \to \infty} \int_\Omega \mathrm{dist}^p(u_j(x), \{\lambda, \mu\})dx = 0.$$

Then, up to extracting a subsequence, either

$$\int_\Omega |u_j(x) - \lambda|^p dx \to 0 \quad \textit{or} \quad \int_\Omega |u_j(x) - \mu|^p dx \to 0 \qquad \textit{as } j \to \infty.$$

When applied to the particular case of $\mathcal{A} = \mathrm{curl}$ one recovers the following results, first established by Ball and James.

Theorem 2.7 (Ball–James 1987 [4]) *Let $\Omega \subset \mathbb{R}^d$ be an open, bounded, and connected set and let $A, B \in \mathbb{R}^\ell \otimes \mathbb{R}^d$ be $(\ell \times d)$-matrices with $\mathrm{rk}(A - B) \geq 2$. Then:*

(A) *If $u \in W^{1,\infty}(\Omega; \mathbb{R}^\ell)$ satisfies the differential inclusion*

$$Du(x) \in \{A, B\} \quad \textit{for a.e. } x \in \Omega, \tag{2.9}$$

then either $Du \equiv A$ or $Du \equiv B$.
(B) *Let $(u_j) \subset W^{1,\infty}(\Omega; \mathbb{R}^\ell)$ be a uniformly norm-bounded sequence of maps such that*

$$\mathrm{dist}(Du_j, \{A, B\}) \to 0 \quad \textit{in measure}.$$

Then, up to extracting a subsequence, either

$$\int_\Omega |Du_j(x) - A| dx \to 0 \quad or \quad \int_\Omega |Du_j(x) - B| dx \to 0 \qquad as\ j \to \infty.$$

The proof of Theorem 2.6 is based on some properties of elliptic operators which are collected in Sect. 2.4. Here and in the next section is will be useful to localize several estimates. To this end, given $\varphi \in C_c^\infty(\mathbb{R}^d)$ and a differential operator $\mathcal{A}$ of order k we introduce the commutator $[\mathcal{A}, \varphi]$ whose action is given by

$$[\mathcal{A}, \varphi](u) = \mathcal{A}(\varphi u) - \varphi \mathcal{A}(u).$$

The following simple lemma will be useful

Lemma 2.8 *Let $\mathcal{A}$ be a differential operator of order k and let $\varphi \in C_c^\infty(\Omega)$, then there exists a differential operator $\mathcal{C}$ of order at most $k - 1$ (but not homogenous)*

$$\mathcal{C} = \sum_{|\alpha| \leq k-1} C_\alpha \partial^\alpha$$

such that for all $u \in L^p(\Omega; R^m)$ there exists a compactly supported $v \in L^p(\Omega; R^N)$, $N = N(m, k, d)$ such that

$$[\mathcal{A}, \varphi]u = \mathcal{C}v. \tag{2.10}$$

Moreover for all $p \in [1, \infty)$,

$$\|v\|_{L^p} \leq C \|u\|_{L^p}.$$

where C depends only on $k, d,$ and φ.

Proof The proof is a simple application of the Leibniz rule. Note that $[\mathcal{A}, \varphi](u)$ is a sum of products of derivatives of u and of φ where the order of the derivative on u is at most $k - 1$, hence it is enough to consider operators of the form

$$\partial^\beta \varphi \partial^\gamma u$$

where $|\gamma| \leq k - 1$. Let us consider $k = 2$ and $\gamma, \beta = 1$ for simplicity, then

$$\partial_1 \varphi \partial_1 u = \partial_1(\partial_1 \varphi u) - \partial_{11} \varphi u$$

so that we can set $v = (v_1, v_2) = (\partial_1 \varphi u, -\partial_{11} \varphi u)$ and

$$\mathcal{C}v = \partial_1 v_1 + v_2.$$

The general case can be obtained in the same way.   □

Note that the above basically states that $[\mathcal{A}, \varphi]$ is an operator of order $(k-1)$. We can now prove Theorem 2.6.

Proof of Theorem 2.6 Let us start with the rigidity statement (A). Note that, up to considering $v - \mu$ we can assume without loss of generality that $\mu = 0$ and $\lambda \notin \Lambda_{\mathcal{A}}$. In this case, by setting $E = \{u = \lambda\}$ we obtain that

$$u = \lambda \mathbf{1}_E.$$

By the $\mathcal{A}$-freeness of u, this implies

$$\mathcal{B}_\lambda(\mathbf{1}_E) = \mathcal{A}u = 0$$

where $\mathcal{B}_\lambda$ was introduced in (2.7). Since $\lambda \notin \Lambda_{\mathcal{A}}$, $\mathcal{B}_\lambda$ is elliptic, hence Lemma 4.4 in Sect. 2.4 implies that $\mathbf{1}_E \in C^\infty(\Omega)$ which forces E to be either empty or the whole Ω.

Let us prove (B). Again we assume $\mu = 0$. Let (ε_j) be an infinitesimal sequence to be chosen below and let us set

$$E_j = \{x \in \Omega : |u_j - \lambda| < \varepsilon_j\}.$$

We split

$$u_j = \lambda \mathbf{1}_{E_j} + r_j,$$

and we estimate

$$\int r_j^p = \int_{E_j} |u_j - \lambda|^p + \int_{E_j^c} u_j^p$$

$$\leq \varepsilon_j^p |\Omega| + \int_{E_j^c \cap \{|u_j| < \varepsilon_j\}} u_j^p + \int_{\Omega \cap \{|u_j| \geq \varepsilon_j\} \cap \{|u_j - \lambda| \geq \varepsilon_j\}} u_j^p.$$

To estimate the last term we note that

$$|u_j| \leq \text{dist}(u_j, \{\lambda, 0\}) + |\lambda|$$

and that on $\{|u_j| \geq \varepsilon_j\} \cap \{|u_j - \lambda| \geq \varepsilon_j\}$, $\text{dist}(u_j, \{\lambda, 0\}) \geq \varepsilon_j$. Hence

$$\int_{\Omega \cap \{|u_j| \geq \varepsilon_j\} \cap \{|u_j - \lambda| \geq \varepsilon_j\}} u_j^p$$

$$\leq 2^{p-1} \int_\Omega \text{dist}^p(u_j, \{\lambda, 0\}) + 2^{p-1} |\lambda|^p |\{\text{dist}(u_j, \{\lambda, 0\}) \geq \varepsilon_j\}|.$$

The first term in the above inequality goes to zero as $j \to \infty$. As for the second one

$$|\{\text{dist}(u_j, \{\lambda, 0\}) \geq \varepsilon_j\}| \leq \frac{1}{\varepsilon_j^p} \int_\Omega \text{dist}^p(u_j, \{\lambda, 0\}) \to 0$$

provided we choose, for instance, $\varepsilon_j = \| \text{dist}(u_j, \{\lambda, 0\}) \|_{L^p}^{1/2}$. Summarizing we have proved that

$$\|r_j\|_{L^p} \to 0.$$

We now let $\varphi \in C_c^\infty(\Omega)$ and we write

$$0 = \varphi \mathcal{A}(u_j) = \mathcal{A}(\varphi u_j) + [\mathcal{A}, \varphi](u_j) = \mathcal{B}_\lambda(\varphi \mathbf{1}_{E_j}) + \mathcal{A}(\varphi r_j) + [\mathcal{A}, \varphi]u_j$$

so that

$$\mathcal{B}_\lambda(\varphi \mathbf{1}_{E_j}) = -\mathcal{A}(\varphi r_j) - [\mathcal{A}, \varphi](u_j)$$

Here $[\mathcal{A}, \varphi]$ is the commutator operator defined in (2.10). By Lemma 2.8, there exists a sequence of functions v_j equi-bounded in L^p and an operator $\mathcal{C}$ of order at most $k - 1$ such that

$$\mathcal{B}_\lambda(\varphi \mathbf{1}_{E_j}) = -\mathcal{A}(\varphi r_j) + \mathcal{C}v_j.$$

By Lemma 4.3 in Sect. 2.4 this implies that $\varphi \mathbf{1}_{E_j}$ is pre-compact in $L^p(\Omega)$. Since φ is arbitrary there exists a set E such that

$$u_j \to \lambda \mathbf{1}_E$$

in L_{loc}^p. By (A) above either $E = \Omega$ or $E = \emptyset$, concluding the proof.

$\square$

2.3 Linear PDE Constrained Measures

In this section we investigate the structure of $\mathcal{A}$-free measures. As we mentioned in the introduction, by suitably choosing $\mathcal{A}$ this allows to recover and improve several known results. For instance:

- If $\mathcal{A} = \text{curl}$ we are studying the structure of gradients which are measures, which is the same that studying BV the space of functions of bounded variation.
- If $\mathcal{A} = \text{curlcurl}$ we are studying the structure of symmetrized gradients which are measures, which corresponds to studying BD the space of functions of bounded deformation.

– When $\mathcal{A} = \mathrm{div}$ we are for instance investigating the structure of stress free tensors or of stationary varifolds. Let us also stress that the study of div free matrix valued measures turned out to have several applications, for instance in proving the converse of Rademacher Theorem in the analysis of metric measures spaces, see [9, Theorem 1.14], and [6–8] and references therein for a survey on this results.

The key point in studying $\mathcal{A}$-free measures consists in understanding their *singular* part. Referring to Appendix A for more details concerning vector valued Radon measures, let us recall that by the Radon-Nykodim theorem we can write every Radon measure $\mu \in \mathcal{M}(\Omega; V)$ as the sum of an absolutely continuous part with respect to the Lebesgue measure and a singular one:

$$\mu = \mu^a + \mu^s = g d \mathcal{L}^d + \frac{d\mu}{d|\mu|} |\mu|^s.$$

Here $g \in L^1(\Omega; V)$, $|\mu|$ is the total variation of μ^1 and $\frac{d\mu}{d|\mu|} \in L^\infty_{|\mu|}(\Omega; V)$ is called the *polar vector* of μ and it represents the *infinitesimal direction* of μ, see Appendix A. Recall also that

$$\frac{d\mu}{d|\mu|}(x) = \lim_{r \to 0} \frac{\mu(B_r(x))}{|\mu|(B_r(x))}$$

and that the limit exists $|\mu|$ almost everywhere.

The question we want to answer in this section is how the differential constraint imposes restriction on the behavior of the singular part of μ. To this end we start by considering the simple case where $\mu = \lambda \nu$ where $\lambda \notin \Lambda_\mathcal{A}$ is a fixed vector that does not belong to the wave cone. In this case the condition $\mathcal{A}\mu = 0$ translates into

$$\mathcal{B}_\lambda \nu = 0$$

and again Weyl's Lemma, Lemma 4.4 implies that ν is smooth. Hence we may expect that at a singular point the polar vector has to belong to the wave cone. This is indeed the content of the next result:

Theorem 3.1 ([9]) *Let $\Omega \subset \mathbb{R}^d$ be an open set, let $\mathcal{A}$ be a k'th-order differential operator, and let μ be an $\mathcal{A}$-free Radon measure on Ω. Then,*

$$\frac{d\mu}{d|\mu|}(x) \in \Lambda_\mathcal{A} \qquad \text{for } |\mu|^s\text{-a.e. } x \in \Omega.$$

---

[1]We are assuming that V is a normed vector space, note however that the results in this section does note depend on the choice of the norm.

Before proving the theorem let us state a few corollaries which immediately follows by the structure of the wave cone when $\mathcal{A} = \text{curl}$, $\mathcal{A} = \text{curlcurl}$ or $\mathcal{A} = \text{div}$. The first one is a new proof of Alberti's rank one theorem concerning the singular part of gradients of BV functions, [1] (see also [18]).

Corollary 3.2 (Alberti's rank-one theorem) *Let $\Omega \subset \mathbb{R}^d$ be an open set and let $u \in BV(\Omega; \mathbb{R}^\ell)$. Then, for $|D^s u|$-almost every $x \in \Omega$, there exist $a(x) \in \mathbb{R}^\ell \setminus \{0\}$, $b(x) \in \mathbb{R}^d \setminus \{0\}$ such that*

$$\frac{dDu}{d|Du|}(x) = a(x) \otimes b(x).$$

The second one is its extension to case of BD functions:

Corollary 3.3 *Let $\Omega \subset \mathbb{R}^d$ be an open set and let $u \in BD(\Omega)$. Then, for $|E^s u|$-almost every $x \in \Omega$, there exist $a(x), b(x) \in \mathbb{R}^d \setminus \{0\}$ such that*

$$\frac{dEu}{d|Eu|}(x) = a(x) \odot b(x).$$

The last one concerns the structure of divergence free measure valued matrixes

Corollary 3.4 *Let $\mu \in \mathcal{M}(\Omega; \mathbb{R}^d \otimes \mathbb{R}^d)$ be a matrix-valued measure such that*

$$\text{div}\, \mu = 0$$

Then,

$$\text{rank}\left(\frac{d\mu}{d|\mu|}(x)\right) \leq d - 1 \qquad \text{for } |\mu|^s\text{-a.e. } x \in \Omega.$$

In order to prove Theorem 3.1 one would argue by contradiction and assume that there is a set E of positive $|\mu|^s$-measure such that the polar vector $\frac{d\mu}{d|\mu|}(x)$ is not in the wave cone $\Lambda_\mathcal{A}$ for every $x \in E$. By, Theorem A.5 and Proposition A.6 in Appendix A, one can find a point $x_0 \in E$ and a sequence $r_j \downarrow 0$ such that

$$\text{w*-}\lim_{j\to\infty} \frac{(T^{x_0,r_j})_\sharp \mu}{|\mu|(B_{r_j}(x_0))} = \text{w*-}\lim_{j\to\infty} \frac{(T^{x_0,r_j})_\sharp \mu^s}{|\mu|^s(B_{r_j}(x_0))} = \lambda \nu,$$

where $(T^{x_0,r_j})_\sharp \mu$ is defined as in (2.26) $\nu \in \text{Tan}(x_0, |\mu|) = \text{Tan}(x_0, |\mu|^s)$ is a non-zero tangent measure and

$$\lambda := \frac{d\mu}{d|\mu|}(x_0) \notin \Lambda_\mathcal{A}.$$

Moreover, one easily checks that

$$0 = \mathcal{A}(\lambda \nu) = \mathcal{B}_\lambda \nu$$

and we are back to the case of the fixed polar discussed above. In this case ν is represented by a smooth function and thus in particular absolutely continuous with respect to the Lebesgue measure. Unfortunately this is not a contradiction with the fact that $\nu \in \mathrm{Tan}(x_0, |\mu|^s)$ since there are examples of singular measures whose tangent spaces contains only absolutely continuous measures, [23]. Hence, to prove the theorem one has to obtain a contradiction "slightly" before passing to the limit. Heuristically we can consider the above discussion as a rigidity statement for $\mathcal{A}$-free measures of the form $\lambda \nu$ and we need to upgrade this rigidity statement into a "stability" statement in order to obtain the desired contradiction.

Proof of Theorem 3.1 As we said we argue by contradiction. Let

$$E = \left\{ x \in \Omega : \frac{d\mu}{d|\mu|}(x) \notin \Lambda_{\mathcal{A}} \right\}.$$

Assume by contradiction that $|\mu|^s(E) > 0$. We now choose a point $x_0 \in E$ and a sequence $r_j \downarrow 0$ such that

(i) $\displaystyle \lim_{j \to \infty} \frac{|\mu|^a(B_{r_j}(x_0))}{|\mu|^s(B_{r_j}(x_0))} = 0$ and $\displaystyle \lim_{j \to \infty} \fint_{B_{r_j}(x_0)} \left| \frac{d\mu}{d|\mu|}(x) - \frac{d\mu}{d|\mu|}(x_0) \right| d|\mu|^s(x) = 0$;

(ii) there exists a positive Radon measure $\nu \in \mathcal{M}_+(\mathbb{R}^d)$ with $\nu \llcorner B_{1/2} \neq 0$ and such that

$$\nu_j := \frac{(T^{x_0, r_j})_\sharp |\mu|^s}{|\mu|^s(B_{r_j}(x_0))} \xrightarrow{\ *\ } \nu;$$

(iii) for the polar vector it holds that

$$\lambda := \frac{d\mu}{d|\mu|}(x_0) \notin \Lambda_{\mathcal{A}}$$

Indeed, (i) holds at $|\mu|^s$-almost every point by (2.25) and (2.24), (ii) follows by the fact that for $|\mu|^s$-almost every $x \in \Omega$ the space of tangent measures $\mathrm{Tan}(|\mu|^s, x)$ to $|\mu|^s$ at x is non-trivial, Theorem A.5, and finally, (iii) follows from Proposition A.6 the assumption $|\mu|^s(E) > 0$.

We now claim that (i)–(iii) above imply that

$$0 \neq \nu \llcorner B_{1/2} \ll \mathcal{L}^d, \tag{2.11}$$

$$\lim_{j \to \infty} |\nu_j - \nu|(B_{1/2}) = 0. \tag{2.12}$$

Before proving (2.11) and (2.12), let us show how to use them to conclude the proof. Recall that $v_j \perp \mathcal{L}^d$ and take Borel sets $E_j \subset B_{1/2}$ with $\mathcal{L}^d(E_j) = 0 = v(E_j)$ and $v_j(E_j) = v_j(B_{1/2})$. Then,

$$v_j(B_{1/2}) = v_j(E_j) \leq |v_j - v|(B_{1/2}) + v(E_j) = |v_j - v|(B_{1/2}) \to 0,$$

thanks to (2.12). Hence, we infer $v(B_{1/2}) = 0$, in contradiction to (2.11). Thus, $|\mu|^s(E) = 0$, concluding the proof of the theorem.

We are thus left to prove (2.11) and (2.12). Let us assume that $x_0 = 0$ and set $T^r := T^{x_0,r}$.

Clearly,

$$\mathcal{A}\big(T^r_\sharp \mu\big) = 0 \qquad \text{in } \mathcal{D}'(B_1; \mathbb{R}^n).$$

Therefore, with v_j defined as in (ii) above and $c_j := |\mu|^s(B_{r_j})^{-1}$,

$$\mathcal{B}_\lambda v_j = \mathcal{A}(\lambda v_j) = \mathcal{A}(\lambda v_j - c_j T^{r_j}_\sharp \mu). \tag{2.13}$$

Let now $\{\varphi_\varepsilon\}_{\varepsilon>0}$ be a compactly supported, smooth, and positive approximation of the identity. By the lower semicontinuity of the total variation,

$$|v_j - v|(B_{1/2}) \leq \liminf_{\varepsilon \to 0} |v_j * \varphi_\varepsilon - v|(B_{1/2}).$$

Thus, for every j we can find $\varepsilon_j \leq 1/j$ such that

$$|v_j - v|(B_{1/2}) \leq |v_j * \varphi_{\varepsilon_j} - v|(B_{1/2}) + \frac{1}{j}. \tag{2.14}$$

We now convolve (2.13) with φ_{ε_j} to get

$$\mathcal{B}_\lambda u_j = \mathcal{A}(r_j), \tag{2.15}$$

where we have set

$$u_j := v_j * \varphi_{\varepsilon_j}, \qquad r_j := \big[\lambda v_j - c_j T^{r_j}_\sharp \mu\big] * \varphi_{\varepsilon_j}.$$

Note that u_j, r_j are smooth, $u_j \geq 0$, and

$$u_j \overset{*}{\rightharpoonup} v. \tag{2.16}$$

Moreover, recalling that $x_0 = 0$ and $c_j = |\mu|^s(B_{r_j})^{-1}$, by the definition of r_j, v_j, λ and standard properties of convolutions, for $\varepsilon_j \leq 1/4$ it holds that

$$\int_{B_{3/4}} |r_j| dx \leq \frac{\left| \lambda T_\sharp^{r_j} |\mu|^s - T_\sharp^{r_j} \mu \right|(B_1)}{|\mu|^s(B_{r_j})}$$

$$\leq \frac{\left| \lambda |\mu|^s - \mu^s \right|(B_{r_j})}{|\mu|^s(B_{r_j})} + \frac{|\mu|^a(B_{r_j})}{|\mu|^s(B_{r_j})}$$

$$= \int_{B_{r_j}} \left| \frac{d\mu}{d|\mu|}(0) - \frac{d\mu}{d|\mu|}(x) \right| d|\mu|^s(x) + \frac{|\mu|^a(B_{r_j})}{|\mu|^s(B_{r_j})}.$$

Hence, by (i) above,

$$\lim_{j \to \infty} \int_{B_{3/4}} |r_j| dx = 0. \tag{2.17}$$

Take a cut-off function $\chi \in C_c^\infty(B_{3/4})$ with $0 \leq \chi \leq 1$ and $\chi \equiv 1$ on $B_{1/2}$. Then, (2.15) implies that

$$\mathcal{B}_\lambda(\chi u_j) = \mathcal{A}(\chi r_j) + [\mathcal{B}_\lambda, \chi](u_j) - [\mathcal{A}, \chi](r_j). \tag{2.18}$$

By Lemma 2.8, there exist a differential operator $\mathcal{C}$ and a compactly supported function v_j such that

$$\|v_j\|_{L^1} \leq C\|u_j\|_{L^1} + C\|r_j\|_{L^1} \leq C$$

so that we can write (2.18) as

$$\mathcal{B}_\lambda(\chi u_j) = \mathcal{A}(\chi r_j) + \mathcal{C}(v_j)$$

We can now use Lemma 4.3, to write χu_j as

$$\chi u_j = f_j + g_j$$

where (f_j) goes to zero in measure and in $\mathcal{D}'(B_1)$ and (g_j) is pre-compact in L^1_{loc}. This would not be a prior enough to deduce the precompactness of u_j in L^1. However we can exploit positivity of u_j to obtain it, indeed, since $\chi u_j \geq 0$, we have that

$$f_j^- := \max\{0, -f_j\} \leq |g_j|.$$

Since family $(g_j)_j$ is precompact in $L^1_{\text{loc}}(\mathbb{R}^d)$ it is equi-integrable. The previous inequality implies then the equi-integrability of (f_j^-). By Lemma 3.5 below this yields $f_j \to 0$ in $L^1_{\text{loc}}(\mathbb{R}^d)$ and thus that the sequence (χu_j) is precompact in

$L^1_{\text{loc}}(\mathbb{R}^d)$. Since also $\chi u_j \overset{*}{\rightharpoonup} \chi v$ by (2.16), we deduce that $\chi v \in L^1(\mathbb{R}^d)$, which implies (2.11). Moreover,

$$\chi u_j \to \chi v \qquad \text{in } L^1_{\text{loc}}(\mathbb{R}^d),$$

which, taking into account (2.14), implies (2.12).   □

Lemma 3.5 *Let* $(f_j) \subset L^1(B_1)$ *be a family of functions such that*

(a) $f_j \overset{*}{\rightharpoonup} 0$ *in* $\mathcal{D}'(B_1)$;
(b) *The negative parts of* f_j *tends to zero in measure, i.e.*

$$\lim_{j\to\infty} \left| \{f_j^- > \varepsilon\} \right| = 0 \qquad \text{for every } \varepsilon > 0;$$

(c) *the sequence of negative parts* (f_j^-) *is equi-integrable,*

$$\lim_{|E|\to 0} \sup_{j\in\mathbb{N}} \int_E f_j^- \, dx = 0.$$

Then, $f_j \to 0$ *in* $L^1_{\text{loc}}(B_1)$.

Proof Let $\varphi \in C_c^\infty(B_1)$, $0 \le \varphi \le 1$. It is enough to show that

$$\lim_{j\to\infty} \int \varphi |f_j| dx = 0. \tag{2.19}$$

We write

$$\int \varphi |f_j| dx = \int \varphi f_j dx + 2 \int \varphi f_j^- dx \le \int \varphi f_j dx + 2 \int f_j^- dx.$$

The first term on the right-hand side goes to 0 as $j \to \infty$ by assumption (a). Thanks to the Vitali convergence theorem, assumptions (b) and (c) further give that also the third term vanishes in the limit. Hence, (2.19) follows.   □

2.3.1 Dimensional Estimates

We conclude this section by mentioning, without proof, some recent extensions of Theorem 3.1. Note that Theorem 3.1 constrains the value of the polar vector of an $\mathcal{A}$-free measure at singular points. This can be rephrased as saying that if E is such that

$$|\mu|(E) > 0 \qquad \text{and} \qquad \mathcal{L}^d(E) = 0$$

then

$$\frac{d\mu}{d|\mu|}(x) \in \Lambda_{\mathcal{A}}$$

for $|\mu|$ almost every $x \in E$. The next natural question is if we can improve the above result when instead of assuming that the d dimensional Lebesgue measure of E is 0, we assume that ℓ-dimensional Hausdorff measure of E is zero (see [19] for an account of Hausdorff measures). To this end we define for $\ell = 1, \dots, d$ we define the *ℓ-dimensional wave cone* as

$$\Lambda_{\mathcal{A}}^{\ell} := \bigcap_{\pi \in \mathrm{Gr}(\ell,d)} \bigcup_{\xi \in \pi \setminus \{0\}} \ker \mathbb{A}(\xi). \tag{2.20}$$

where $\mathrm{Gr}(\ell, d)$ is the Grassmanian of ℓ dimensional spaces in $\mathbb{R}^d$. Note that, by the very definition of $\Lambda_{\mathcal{A}}^{\ell}$, we have the following inclusions:

$$\Lambda_{\mathcal{A}}^1 = \bigcap_{\xi \in \mathbb{R}^d \setminus \{0\}} \ker \mathbb{A}(\xi) \subset \Lambda_{\mathcal{A}}^j \subset \Lambda_{\mathcal{A}}^{\ell} \subset \Lambda_{\mathcal{A}}^d = \Lambda_{\mathcal{A}}, \qquad 1 \le j \le \ell \le d.$$

$$\tag{2.21}$$

The main result of [3] refines Theorem 3.1 by showing that the polar vector is more constrained at point where μ is singular with respect to the ℓ dimensional measure:

Theorem 3.6 ([3]) *Let $\Omega \subset \mathbb{R}^d$ be open, let $\mathcal{A}$ be a differential operator of order k. Let $\mu \in \mathcal{M}(U; \mathbb{R}^m)$ be an $\mathcal{A}$-free measure on Ω. If $E \subset \Omega$ is a Borel set with $\mathcal{H}^{\ell}(E) = 0$ for some $\ell \in \{0, \dots, d\}$, then*

$$\frac{d\mu}{d|\mu|}(x) \in \Lambda_{\mathcal{A}}^{\ell} \qquad \text{for } |\mu|\text{-a.e. } x \in E. \tag{2.22}$$

Note it might happen that $\Lambda_{\mathcal{A}}^{\ell} = \{0\}$ for some $\ell < d$. In this case every $\mathcal{A}$-free measure is absolutely continuous with respect to $\mathcal{H}^{\ell}$. For instance since one easily verifies that $\Lambda_{\mathrm{curl}}^{d-1} = \{0\}$, one recovers by the above theorem the well known fact that

$$|Du| \ll \mathcal{H}^{d-1}$$

for all functions $u \in BV$. Furthermore in Theorem 3.6 one can replace $\mathcal{H}^{\ell}$ with the smaller *integral geometric measure* $\mathcal{I}^{\ell}$. Since there are sets with $\mathcal{H}^{\ell}(E) > 0$ but $\mathcal{I}^{\ell}(E) = 0$ the improvement is not trivial. In particular, by using the Besicovitch Federer rectifiability criterion, one can use this refined version to obtain various rectifiability result for $\mathcal{A}$-free measures. We refer the reader to the paper [3] for a discussion of these results and of their applications and also of some open problems.

Here we mention only the following result which is an interesting instance on the interplay between a differential constraint and a pointwise one.

Proposition 3.7 *Let (P, σ) be a ℓ dimensional stationary varifold (i.e. a couple satisfying (2.2) and (2.3)) then $\mu \ll \mathcal{H}^\ell$.*

Proof Assume there exists E such that $\mathcal{H}^\ell(E) = 0$ but $\sigma(E) = 0$. By Theorem 3.6,

$$P(x) \in \Lambda_{\text{div}}^\ell$$

for almost all x. On the other hand, by a direct computation

$$\Lambda_{\text{div}}^\ell = \{M : \text{rk } M \leq \ell - 1\}.$$

On the other hand, since P is an average of orthogonal projections on ℓ planes, it is easy to check that

$$\text{rk } P \geq \ell,$$

a contradiction. $\square$

2.4   Elliptic Operators and Harmonic Analysis Tools

In this section we state and prove some results concerning elliptic operators. Recall that an homogenous differential operator

$$\mathcal{B} = \sum_{|\alpha|=k} B_\alpha \partial^\alpha : C_c^\infty(\Omega; V) \to C_c^\infty(\Omega : W)$$

is said to be elliptic if its symbol

$$\mathbb{B}(\xi) = \sum_{|\alpha|=k} B_\alpha \xi^\alpha \in \text{Lin}(V, W)$$

is injective. Since V, W are finite dimensional space, in the sequel we will identify them with $\mathbb{R}^m$ and $\mathbb{R}^n$ respectively and endow the latter with the classical Euclidean structure. Note that the identification is not canonical but it will not affect the qualitative behavior of the estimates (but it will affect the numerical value of the constants).

We also recall some basic facts of harmonic and Fourier analysis which will be used in the sequel, see [15, 16] for a more detailed account. We will denote by $\mathcal{S}(\mathbb{R}^d; \mathbb{R}^N)$ the space of (vector valued) Schwartz functions with values in $\mathbb{R}^N$ and by $\mathcal{S}'(\mathbb{R}^r; \mathbb{R}^N)$ the space of tempered distribution (which is the topological dual of

$\mathcal{S}(\mathbb{R}^d; \mathbb{R}^N))$. Given a function $u \in \mathcal{S}(\mathbb{R}^r; \mathbb{R}^N)$ its Fourier transform is defined as

$$\hat{u}(\xi) = \mathcal{F}(u)(\xi) = \int u(x)e^{-\mathrm{i}x\cdot\xi}\,dx.$$

By classical duality the Fourier transform can be extended to $\mathcal{S}(\mathbb{R}^r; \mathbb{R}^n)$. Moreover the Fourier is invertible on $\mathcal{S}'(\mathbb{R}^r; \mathbb{R}^N)$ and its inverse is given by

$$\mathcal{F}^{-1}(v)(\xi) = \frac{1}{(2\pi)^d} \int v(\xi)e^{\mathrm{i}x\cdot\xi}\,d\xi.$$

One of the key properties of the Fourier transform is that it transforms differential operators into multiplication operators, indeed

$$\mathcal{F}(\mathcal{A}u)(\xi) = (i)^k \mathbb{A}(\xi)\hat{u}(\xi).$$

Given a bounded and smooth function $a \in C^\infty(\mathbb{R}^d, \mathbb{R}^n \otimes \mathbb{R}^m)$ with at most polynomial growth we define the operator $T_a : \mathcal{S}'(\mathbb{R}^r; \mathbb{R}^m) \to \mathcal{S}'(\mathbb{R}^r; \mathbb{R}^n)$ via

$$T_a(u) = \mathcal{F}^{-1}(a(\xi)\hat{u}(\xi)).$$

Note that thanks to the growth assumption on m, the operator indeed maps $\mathcal{S}'(\mathbb{R}^r; \mathbb{R}^m)$ into $\mathcal{S}'(\mathbb{R}^r; \mathbb{R}^n)$.

In the previous section we had to deal with equation of the form

$$\mathcal{B}u = \mathcal{A}r + \mathcal{C}v.$$

for some functions where $\mathcal{B}$ is an elliptic differential operator of order k, $\mathcal{A}$ is an elliptic operator of order k and $\mathcal{C}$ at most $k - 1$. Though $\mathcal{B}$ is (formally) injective we do not know it to be surjective.[2] Formally, in order to solve this problem we can apply to both side by $\mathcal{B}^*$ (the transpose of $\mathcal{B}$) to obtain

$$(\mathrm{Id} + \mathcal{B}^*\mathcal{B})u = \mathcal{B}^*\mathcal{A}r + \mathcal{B}^*\mathcal{C}v + u$$

and invert $\mathcal{B}^*\mathcal{B}$. Actually for technical reasons it is better to add u to both sides and to invert $(\mathrm{Id} + \mathcal{B}^*\mathcal{B})$ to obtain

$$u = (\mathrm{Id} + \mathcal{B}^*\mathcal{B})^{-1}\mathcal{B}^*\mathcal{A}r + (\mathrm{Id} + \mathcal{B}^*\mathcal{B})^{-1}\mathcal{B}^*\mathcal{C}v + (\mathrm{Id} + \mathcal{B}^*\mathcal{B})^{-1}u.$$

The next result formalize this procedure and contains some estimates.

---

[2]This is due to the fact that we are dealing with vector valued operators, for instance the equation $\mathcal{D}u = f$ can not be solved unless f is curl-free.

Lemma 4.1 *Let B be a homogenous elliptic operator of order k.*

(i) *If we denote by B^* the adjoint of B, then $\mathrm{Id} + B^* B$ is invertible as a map from $S'(\mathbb{R}^d)$ into itself.*

(ii) *If A is an operator of order k then the operator $(\mathrm{Id} + B^* B)^{-1} B^* A$ extends to a bounded operator from L^p to L^p for $p \in (1, \infty)$ and from $L^1 \to L^{1,\infty}$.*

(iii) *If C is an operator of order ℓ with $\ell \leq k - 1$ then the operator $(\mathrm{Id} + B^* B)^{-1} B^* C$ extends to a compact operator from L^1_c to L^1_{loc}.*

Before proving it we recall the following theorem concerning the mapping properties of the operator T_m.

Theorem 4.2 (Hörmander-Mikhlin [15, Theorem 5.2.7]) *Let $m \in C^\infty(\mathbb{R}^d, \mathbb{R}^n \otimes \mathbb{R}^m)$ and assume that for all multi-indexes β,*

$$|\partial^\beta m(\xi)| \leq C_\beta |\xi|^{-|\beta|}.$$

Then for $p \in (1, \infty)$ the operator T_m can be extended to a bounded operator from L^p to L^p. Moreover T_m maps L^1 into $L^{1,\infty}$ which is the set of functions such that

$$\|u\|_{L^{1,\infty}} := \sup_{\lambda > 0} \lambda |\{|u| > \lambda\}| < +\infty.$$

Furthermore

$$\|T_m u\|_{L^{1,\infty}} \leq C \|u\|_{L^1}.$$

Proof of Lemma 4.1 As for (i) is easy to see that the symbol of $\mathrm{Id} + B^* B$ is given by $\mathrm{Id} + \mathbb{B}(\xi)^T \mathbb{B}(\xi)$ and that this matrix is invertible. It is then an easy computation to check that the inverse of $\mathrm{Id} + B^* B$ is given by T_a where[3]

$$a(\xi) = (\mathrm{Id} + \mathbb{B}(\xi)^T \mathbb{B}(\xi))^{-1}.$$

As for (ii) the operator $(\mathrm{Id} + B^* B)^{-1} B^* A$ has a symbol given by

$$(\mathrm{Id} + \mathbb{B}(\xi)^T \mathbb{B}(\xi))^{-1} \mathbb{B}(\xi)^T A(\xi)$$

and this satisfies the assumption of the Hörmander-Mikhlin theorem, Theorem 4.2. Here is were injectivity of $\mathbb{B}(\xi)$ plays a role (and we invite the reader to check it). As for (iii) we can assume that C is homogenous of degree ℓ (by applying the result to each homogeneous component). In this case we write

$$(\mathrm{Id} + B^* B)^{-1} B^* C = (\mathrm{Id} + B^* B)^{-1} B^* C (\mathrm{Id} - \Delta)^{k-\ell} (\mathrm{Id} - \Delta)^{\ell - k}$$

---

[3]The reason to add the identity is precisely to obtain that a is bounded and smooth. Indeed the symbol of $(B^* B)^{-1}$ is $(\mathbb{B}(\xi)^T \mathbb{B}(\xi))^{-1}$ which is not bounded close to the origin. This would introduce a few technical complications.

where for every $s \in \mathbb{R}$ we define

$$(\mathrm{Id} - \Delta)^s$$

as the operator whose Fourier symbol is $(1 + |\xi|^2)^s$. Since $\mathcal{C}$ is a homogeneous operator of order ℓ, it is easy to check that the symbol of the operator $(\mathrm{Id} + \mathcal{B}^*\mathcal{B})^{-1}\mathcal{B}^*\mathcal{C}(\mathrm{Id} - \Delta)^{k-\ell}$ satisfies the assumption of the Hörmander-Mikhlin Theorem and thus maps L^p into itself. On the other hand, by [16, Section 6.1.2], the operator $(\mathrm{Id} - \Delta)^{\ell-k}$ is given by the convolution with the kernel $K(y) \in L^q$ for a suitable $q = q(k, \ell, d) > 1$. Hence the map $u \mapsto (\mathrm{Id} - \Delta)^{\ell-k}$ is compact from L_c^1 to L^q. By combining the above information and recalling that the composition of a compact operator with a bounded one is compact, we conclude the proof, see [9, Lemma 2.1] for more details. □

A simple consequence is the following lemma which has been used in the proof of Theorems 2.6 and 3.1.

Lemma 4.3 *Let $\mathcal{B}$ be an homogeneous elliptic operator of order k, $\mathcal{A}$ be an operator of order k and $\mathcal{C}$ be an operator of order at most $k - 1$. Assume that $u_j, r_j, v_j \in L^p$, $p \in [1, +\infty)$ are uniformly bounded sequence of compactly supported functions such that*

$$\mathcal{B}u_j = \mathcal{A}r_j + \mathcal{C}v_j.$$

and that

$$\|r_j\|_{L^p} \to 0.$$

Then we can write

$$u_j = f_j + g_j$$

where g_j is pre-compact in L_{loc}^p and

(i) *If $p \in (1, +\infty)$, $\|f_j\|_{L^p} \to 0$.*
(ii) *If $p = 1$, f_j goes to zero in measure and in distribution.*

Proof We write

$$u_j = (\mathrm{Id} + \mathcal{B}^*\mathcal{B})^{-1}\mathcal{B}^*\mathcal{A}r_j + (\mathrm{Id} + \mathcal{B}^*\mathcal{B})^{-1}\mathcal{B}^*\mathcal{C}v_j + (\mathrm{Id} + \mathcal{B}^*\mathcal{B})^{-1}u_j.$$

and we set

$$f_j = (\mathrm{Id} + \mathcal{B}^*\mathcal{B})^{-1}\mathcal{B}^*\mathcal{A}r_j \qquad g_j = (\mathrm{Id} + \mathcal{B}^*\mathcal{B})^{-1}\mathcal{B}^*\mathcal{C}v_j + (\mathrm{Id} + \mathcal{B}^*\mathcal{B})^{-1}u_j.$$

By Lemma 4.1 (iii) g_j is pre-compact in L^p_{loc}. Moreover by Lemma 4.1 (iii), if $p \in (1, \infty)$,

$$\|f_j\|_{L^p} \leq C \|r_j\|_{L^p} \to 0$$

while of $p = 1$,

$$\|f_j\|_{L^{1,\infty}} \leq C \|r_j\|_{L^1} \to 0$$

which implies the convergence in measure. The convergence to zero in distribution of f_j follows simply by noticing that

$$\langle f_j, \varphi \rangle = \langle r_j, \left((\mathrm{Id} + B^* B)^{-1} B^* A\right)^* \varphi \rangle \to 0$$

for all $\varphi \in C_c^\infty(\mathbb{R}^d)$.   □

We conclude this section with the following Lemma, used in the proof of Theorem 2.6.

Lemma 4.4 (Weyl) *Let Ω be an open set and let B be an elliptic operator of order k and let $u \in \mathcal{D}'(\Omega)$ be a solution of*

$$Bu = 0$$

then $u \in C^\infty(\Omega)$.

We will prove Lemma 4.4 under the additional assumption that $u \in L^2(\Omega)$. To this end, recall that a tempered distribution u belongs to the Sobolev space $H^s(\mathbb{R}^d)$ if

$$(\mathrm{Id} - \Delta)^{s/2} u \in L^2(\mathbb{R}^d)$$

which, thanks to Plancherel theorem, is equivalent to $(1 + |\xi|^2)^{s/2} \hat{u}(\xi) \in L^2(\mathbb{R}^d)$. Given an open set Ω we say that a distribution $u \in \mathcal{D}'(\Omega)$ is in H^k_{loc} of Ω if $\varphi u \in H^k(\mathbb{R}^d)$ for all $\varphi \in C_c^\infty(\Omega)$. Moreover

$$\bigcap_s H^s_{\mathrm{loc}}(\Omega) = C^\infty(\Omega).$$

It is also easy to see that if A is an operator of order at most ℓ and $u \in H^k(\mathbb{R}^d)$ then $Au \in H^{k-\ell}(\mathbb{R}^d)$, see [13, Chapter 9] for more details.

Proof of Lemma 4.4 Let $\varphi \in C_c^\infty(\Omega)$, then by Lemma 2.8 we have an operator C of order at most $k - 1$ and a function $v \in L^2$ such that

$$B(\varphi u) = [B, \varphi](u) = Cv$$

We now apply $(\mathrm{Id} - \Delta)$ on both side of the of the above inequality to get

$$\mathcal{B}((\mathrm{Id} - \Delta)\varphi u) = (\mathrm{Id} - \Delta)\mathcal{C}v.$$

Since $(\mathrm{Id} - \Delta)\mathcal{C}$ is an operator of order at most k and $v \in L^2$ we can apply Lemma 4.1 (ii) to deduce that $(\mathrm{Id} - \Delta)\varphi u \in L^2$, i.e. $u \in H^1_{\mathrm{loc}}(\Omega)$. By iterating this argument one deduces that

$$u \in \bigcap_s H^s_{\mathrm{loc}}(\Omega) = C^\infty(\Omega).$$

$\square$

Appendix A: A Brief Course on Radon Measures

A.1 Basic Facts

In this section we recall some basic facts about Radon measures. Recall that a Borel positive measure σ on $\Omega \subset \mathbb{R}^d$ is said to be Radon if

$$\sigma(K) < \infty \qquad \text{for all compact } K \subset \Omega.$$

Given a finite dimensional normed space $(V, |\cdot|)$, we say that μ is a V valued Radon measure, $\mu \in \mathcal{M}(\Omega; V)$ if there exists a positive valued Radon measure ν and a function $g \in L^1_{\nu,\mathrm{loc}}(\Omega; V)$ such that

$$\mu = g\sigma \tag{2.23}$$

and for all set $E \Subset \Omega$ we define

$$\mu(E) = \int_E g d\sigma \in V.$$

Note that we can identify μ as a continuous operator[4] on $C^0_c(\Omega; V^*)$ (where V^* is the dual of V) via

$$\langle \mu, \varphi \rangle = \int \langle g, \varphi \rangle_{V,V^*} d\sigma.$$

---

[4] Here $C^0_c(\Omega; V^*)$ is endowed withe the usual inductive limit topology. A sequence (φ_j) converges to a function φ in this topology if and only if all the support of the sequence are contained in a given compact set and φ_j converges to φ uniformly.

Actually one can prove that all linear and continuous operators on $C_c^0(\Omega; V^*)$ are of this form, [10, Theorem 1.38]. By identifying Radon measures with linear operators we have a natural notion of weak* convergence for Radon measures:

Definition A.1 We say that a sequence of Radon measure (μ_j) weakly* converges to μ if

$$\langle \mu_j, \varphi \rangle \to \langle \mu, \varphi \rangle$$

for all $\varphi \in C_c^0(\Omega, V^*)$.

We thus have the following

Theorem A.2 (Banach-Alaouglu) *Let (μ_j) be a sequence of Radon measure such that*

$$\sup_j |\mu_j|(K) < +\infty$$

for all compact K. Then there exists a subsequence $(\mu_{j'})$ and a Radon measure μ such that $\mu_{j'} \overset{*}{\rightharpoonup} \mu$.

Note that g and σ in (2.23) are not uniquely defined, however there exists a canonic form to associate to μ a positive Radon measure and a function g. We start with the following:

Definition A.3 Given finite dimensional normed space V and a measure $\mu \in \mathcal{M}(\Omega; V)$ we define $|\mu|$ as the unique positive Radon measure such that for all open subset of $U \subset \Omega$

$$|\mu|(U) = \sup \left\{ \langle \mu, \varphi \rangle : \varphi \in C_c^0(U, V), |\varphi(x)|_{V^*} \le 1 \right\}.$$

where $| \cdot |_{V^*}$ is the dual norm on V^*.

Given a vector valued Radon measure μ and positive Radon measures σ we say that μ is absolutely continuous with respect to ν ($\mu \ll \sigma$) if for all set E

$$\nu(E) = 0 \Rightarrow |\mu|(E) = 0.$$

Moreover we say that μ is singular with respect to σ ($\mu \perp \sigma$) if there exists a set E such that $|\mu|(E) = 0$ and $\sigma(E^c) = 0$.

If μ is a vector valued Radon measure and σ is a positive Radon measure, the Radon-Nikodym theorem [10, Theorem 1.31], ensures that we can always uniquely decompose μ as

$$\mu = \mu^a + \mu^s$$

where μ^a is absolutely continuous with respect to σ and μ^s is singular. Moreover

$$\mu^a = g\sigma$$

where $g \in L^1_{\sigma,\mathrm{loc}}(\Omega; V)$ can be obtained as

$$g(x) = \lim_{r \to 0} \frac{\mu(B_r(x))}{\sigma(B_r(x))}$$

and the limit exists at ν almost all point, [10, Theorem 1.32]. Usually g is denoted by $\frac{d\mu}{d\nu}$. Note that this implies that if $\mu \perp \sigma$ then at σ almost all points

$$\lim_{r \to 0} \frac{\mu(B_r(x))}{\sigma(B_r(x))} = 0. \tag{2.24}$$

The above discussion implies that there exists a unique function $\frac{d\mu}{d|\mu|} \in L^\infty_{|\mu|}(\Omega; V)$ with $\left|\frac{d\mu}{d|\mu|}(x)\right| = 1$ for $|\mu|$ almost all x such that

$$\mu = \frac{d\mu}{d|\mu|}|\mu|.$$

Moreover, at $|\mu|$ almost all points

$$\frac{d\mu}{d|\mu|}(x) = \lim_{r \to 0} \frac{\mu(B_r(x))}{|\mu|(B_r(x))} \quad \text{and} \quad \lim_{r \to 0} \fint_{B_r(x)} \left|\frac{d\mu}{d|\mu|}(x) - \frac{d\mu}{d|\mu|}(y)\right| d|\mu|(y) = 0. \tag{2.25}$$

A.2 Tangent Measures

It is often important to understand the "infinitesimal" behavior of a measure around a "typical" point x. To this end one would like to perform a blow up analysis around x. In order to do that given a Radon measure $\mu \in \mathcal{M}(\Omega, V)$ on Ω and a ball $B_r(x_0) \subset \Omega$ we define the measure $(T^{x_0,r})_\sharp \mu \in \mathcal{M}(B_1, V)$ as

$$(T^{x_0,r})_\sharp \mu(E) = \mu(x_0 + rE) \tag{2.26}$$

and we set

$$\nu_r = \frac{(T^{x_0,r})_\sharp \mu}{|\mu|(B_r(x_0))}$$

so that $|\nu|(B_1) = 1$. The idea is that the limit as $r \to 0$ of ν_r will "capture" the behavior of μ around x_0. This leads to the notion of tangent measures, introduced by Preiss in [24].

Definition A.4 We say that v is a tangent measure to μ at x_0, $v \in \text{Tan}(\mu, x_0)$ if there exists a sequence $r_j \to 0$ such that

$$v_{r_j} \overset{*}{\rightharpoonup} v.$$

Note that the sequence (v_{r_j}) is always equibounded and thus pre compact with respect to the weak* topology, however one has to be sure that (at least almost every where) the limit is non trivial. This is the content of the next result, where we also collect a few useful properties of tangent measures, see [2] and [24] for a proof.

Theorem A.5 Let $\mu \in \mathcal{M}(\Omega, V)$, then for $|\mu|$ almost every x_0, there exists $v \in \text{Tan}(\mu, x_0)$ such that $v \llcorner B_{1/2} \neq 0$. In particular $\text{Tan}(\mu, x_0) \setminus \{0\}$ is not empty.

We also need some simple locality property of the tangent space, which easily follows from the Lebesgue differentiation theorem.

Proposition A.6 Let $\mu \in \mathcal{M}(\Omega, V)$ and let σ be a positive Radon measure, then for σ almost every x_0,

$$\text{Tan}(\mu, x_0) = \frac{d\mu}{d\sigma}(x_0)\text{Tan}(\sigma, x_0).$$

In particular

$$\text{Tan}(\mu, x_0) = \frac{d\mu}{d|\mu|}(x_0)\text{Tan}(|\mu|, x_0).$$

and, if we denote by $|\mu|^s$ the singular part of $|\mu|$ with respect to σ, for $|\mu|^s$ almost every x_0 it holds

$$\text{Tan}(\mu, x_0) = \frac{d\mu}{d|\mu|}(x_0)\text{Tan}(|\mu|^s, x_0).$$

Acknowledgments The author would like to deeply acknowledge Isaac Neal and Matteo Focardi for a very careful reading of a preliminary version of these notes and for their comments and corrections.

References

1. G. Alberti, Rank one property for derivatives of functions with bounded variation. Proc. R. Soc. Edinb. Sect. A **123**, 239–274 (1993)
2. L. Ambrosio, N. Fusco, D. Pallara, *Functions of Bounded Variation and Free Discontinuity Problems*. Oxford Mathematical Monographs (The Clarendon Press/Oxford University Press, New York, 2000). MR 1857292 (2003a:49002)

3. A. Arroyo-Rabasa, G. De Philippis, J. Hirsch, F. Rindler, Dimensional estimates and rectifiability for measures satisfying linear PDE constraints. Geom. Funct. Anal. **29**(3), 639–658 (2019). MR 3962875
4. J.M. Ball, R.D. James, Fine phase mixtures as minimizers of energy. Arch. Ration. Mech. Anal. **100**, 13–52 (1987)
5. E. Chiodaroli, E. Feireisl, O. Kreml, E. Wiedemann, $\mathcal{A}$-free rigidity and applications to the compressible Euler system. Ann. Mat. Pura Appl. (4) **196**(4), 1557–1572 (2017). MR 3673680
6. G. De Philippis, *On the Singular Part of Measures Constrained by Linear PDEs and Applications*. European Congress of Mathematics (European Mathematical Society, Zürich, 2018), pp. 833–845. MR 3890453
7. G. De Philippis, A. Marchese, F. Rindler, On a conjecture of Cheeger, in *Measure Theory in Non-smooth Spaces*, ed. by N. Gigli (De Gruyter Open, Warsaw/Berlin, 2017), pp. 145–155
8. G. De Philippis, L. Palmieri, F. Rindler, On the two-state problem for general differential operators. Nonlinear Anal. Part B **177**, 387–396 (2018). MR 3886580
9. G. De Philippis, F. Rindler, On the structure of $\mathcal{A}$-free measures and applications. Ann. Math. **184**, 1017–1039 (2016)
10. L. Evans, R.F. Gariepy, *Measure Theory and Fine Properties of Functions*, revised ed. Textbooks in Mathematics (CRC Press, Boca Raton, 2015). MR 3409135
11. L.C. Evans, *Weak Convergence Methods for Nonlinear Partial Differential Equations*. CBMS Regional Conference Series in Mathematics, vol. 74 (1990)
12. H. Federer, *Geometric Measure Theory*. Die Grundlehren der mathematischen Wissenschaften, Band 153 (Springer, New York, 1969). MR MR0257325 (41 #1976)
13. G.B. Folland, *Real Analysis*. Pure and Applied Mathematics (New York). A Wiley-Interscience Publication, 2nd edn. (John Wiley & Sons, New York, 1999)
14. I. Fonseca, S. Müller, $\mathcal{A}$-quasiconvexity, lower semicontinuity, and Young measures. SIAM J. Math. Anal. **30**(6), 1355–1390 (1999)
15. L. Grafakos, *Classical Fourier Analysis*. Graduate Texts in Mathematics, vol. 249, 3rd edn. (Springer, New York, 2014)
16. L. Grafakos, *Modern Fourier Analysis*. Graduate Texts in Mathematics, vol. 250, 3d edn. (Springer, New York, 2014). MR 3243741
17. B. Kirchheim, *Rigidity and Geometry of Microstructures*. Lecture Notes 16 (Max-Planck-Institut für Mathematik in den Naturwissenschaften, Leipzig, 2003)
18. A. Massaccesi, D. Vittone, An elementary proof of the rank-one theorem for BV functions. J. Eur. Math. Soc. **21**(10), 3255–3258 (2019)
19. P. Mattila, *Geometry of Sets and Measures in Euclidean Spaces*, vol. 44 (Cambridge University Press, Cambridge, 1995)
20. S. Müller, Variational models for microstructure and phase transitions, in *Calculus of Variations and Geometric Evolution Problems* (Cetraro, 1996), vol. 1713 (Springer, Berlin, 1999), pp. 85–210
21. F. Murat, Compacité par compensation. Ann. Sc. Norm. Sup. Pisa Cl. Sci. **5**, 489–507 (1978)
22. F. Murat, *Compacité par compensation. II*, Proceedings of the International Meeting on Recent Methods in Nonlinear Analysis, Rome, 1978 (Pitagora Editrice Bologna, 1979), pp. 245–256
23. T. O'Neil, A measure with a large set of tangent measures. Proc. Am. Math. Soc. **123**, 2217–2220 (1995)
24. D. Preiss, Geometry of measures in $\mathbf{R}^n$: distribution, rectifiability, and densities. Ann. Math. **125**, 537–643 (1987)
25. L. Simon, *Lectures on Geometric Measure Theory*, Proceedings of the Centre for Mathematical Analysis, Australian National University, vol. 3 (Australian National University Centre for Mathematical Analysis, Canberra, 1983). MR MR756417 (87a:49001)
26. L. Tartar, Compensated compactness and applications to partial differential equations, in *Nonlinear Analysis and Mechanics: Heriot-Watt Symposium*, Vol. IV. Research Notes in Mathematics, vol. 39 (Pitman, 1979), pp. 136–212

Chapter 3
Regularity of Free Boundaries in Obstacle Problems

Xavier Ros-Oton

Abstract Free boundary problems are those described by PDE that exhibit a priori unknown (free) interfaces or boundaries. Such type of problems appear in Physics, Geometry, Probability, Biology, or Finance, and the study of solutions and free boundaries uses methods from PDE, Calculus of Variations, and Geometric Measure Theory. The main mathematical challenge is to understand the regularity of free boundaries. The Stefan problem and the obstacle problem are the most classical and motivating examples in the study of free boundary problems. A milestone in this context is the classical work of Caffarelli, in which he established for the first time the regularity of free boundaries in the obstacle problem, outside a certain set of singular points. This is one of the main results for which he got the Wolf Prize in 2012 and the Shaw Prize in 2018.

The goal of these notes is to introduce the obstacle problem, prove some of the main known results in this context, and give an overview of more recent research on this topic.

3.1 Introduction

The most basic mathematical question in PDEs is that of regularity:

Are all solutions to a given PDE smooth, or may they have singularities?

X. Ros-Oton (✉)
Institut für Mathematik, Universität Zürich, Zürich, Switzerland

ICREA, Barcelona, Spain

Departament de Matemàtiques i Informàtica, Universitat de Barcelona, Barcelona, Spain
e-mail: xros@ub.edu

© The Author(s), under exclusive license to Springer Nature Switzerland AG 2021
M. Focardi, E. Spadaro (eds.), *Geometric Measure Theory and Free Boundary Problems*, Lecture Notes in Mathematics 2284,
https://doi.org/10.1007/978-3-030-65799-4_3

In some cases, all solutions are smooth and no singularities appear (as in Hilbert's XIXth problem [11, 26, 27]). However, in many other cases singularities do appear, and then the main issue becomes the structure of the singular set.

In these notes we will study a special type of elliptic PDE: a *free boundary problem*. In this kind of problems we are not only interested in the regularity of a solution u, but also in the study of an a priori unknown interphase Γ (the free boundary).

As explained later, there is a wide variety of problems in physics, industry, biology, finance, and other areas which can be described by PDEs that exhibit free boundaries. Many of such problems can be written as variational inequalities, for which the solution is obtained by minimizing a constrained energy functional. And the most important and canonical example is the classical *obstacle problem*.

3.1.1 The Obstacle Problem

Given a smooth function φ, the obstacle problem is the following:

$$\text{minimize} \quad \frac{1}{2}\int_\Omega |\nabla v|^2 dx \quad \text{among all functions } v \geq \varphi. \tag{3.1}$$

Here, the minimization is subject to boundary conditions $v|_{\partial\Omega} = g$.

The interpretation of such problem is clear: One looks for the least energy function v, but the set of admissible functions consists only of functions that are above a certain "obstacle" φ.

In 2D, one can think of the solution v as a "membrane" which is elastic and is constrained to be above φ (see Fig. 3.1).

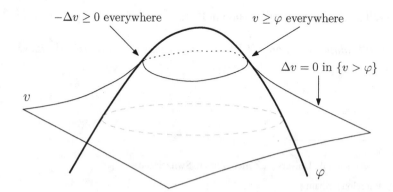

Fig. 3.1 The function v minimizers the Dirichlet energy among all functions with the same boundary values situated above the obstacle

The Euler–Lagrange equation of the minimization problem is the following:

$$\begin{cases} v \geq \varphi \text{ in } \Omega \\ \Delta v \leq 0 \text{ in } \Omega \\ \Delta v = 0 \text{ in the set } \{v > \varphi\}, \end{cases} \tag{3.2}$$

together with the boundary conditions $v|_{\partial\Omega} = g$.

Indeed, notice that if we denote $\mathcal{E}(v) = \frac{1}{2}\int_{\Omega} |\nabla v|^2 dx$, then we will have

$$\mathcal{E}(v + \varepsilon\eta) \geq \mathcal{E}(v) \quad \text{for every } \varepsilon \geq 0 \text{ and } \eta \geq 0, \; \eta \in C_c^{\infty}(\Omega),$$

which yields $\Delta v \leq 0$ in Ω. That is, we can perturb v with *nonnegative* functions $(\varepsilon\eta)$ and we always get admissible functions $(v+\varepsilon\eta)$. However, due to the constraint $v \geq \varphi$, we cannot perturb v with negative functions in all of Ω, but only in the set $\{v > \varphi\}$. This is why we get $\Delta v \leq 0$ *everywhere* in Ω, but $\Delta v = 0$ *only* in $\{v > \varphi\}$. (We will show later that any minimizer v of (3.1) is continuous, so that $\{v > \varphi\}$ is open.)

Alternatively, we may consider $u := v - \varphi$, and the problem is equivalent to

$$\begin{cases} u \geq 0 \text{ in } \Omega \\ \Delta u \leq f \text{ in } \Omega \\ \Delta u = f \text{ in the set } \{u > 0\}, \end{cases} \tag{3.3}$$

where $f := -\Delta\varphi$.

Such solution u can be obtained as follows:

$$\text{minimize} \quad \int_{\Omega} \left\{ \frac{1}{2}|\nabla u|^2 + fu \right\} dx \quad \text{among all functions } u \geq 0. \tag{3.4}$$

In other words, we can make the *obstacle* just *zero*, by adding a *right-hand side* f. Here, the minimization is subject to the boundary conditions $u|_{\partial\Omega} = \tilde{g}$, with $\tilde{g} := g - \varphi$.

3.1.2 On the Euler–Lagrange Equations

As said above, the Euler–Lagrange equations of the minimization problem (3.1) are:

(i) $v \geq \varphi$ in Ω (v is *above* the *obstacle*).
(ii) $\Delta v \leq 0$ in Ω (v is a *supersolution*).
(iii) $\Delta v = 0$ in $\{v > \varphi\}$ (v is *harmonic* where it *does not touch* the obstacle).

These are inequalities, rather than a single PDE. Alternatively, one can write also the Euler–Lagrange equations in the following way

$$\min\{-\Delta v,\; v - \varphi\} = 0 \quad \text{in} \quad \Omega. \tag{3.5}$$

Of course, the same can be done for the equivalent problem (3.3). In that case, moreover, the minimization problem (3.4) is equivalent to

$$\text{minimize} \qquad \int_{\Omega} \left\{ \frac{1}{2} |\nabla u|^2 + f u^+ \right\} dx, \tag{3.6}$$

where $u^+ = \max\{u, 0\}$. In this way, we can see the problem not as a constrained minimization but as a minimization problem with a non-smooth term u^+ in the functional. The Euler–Lagrange equation for this functional is then

$$\Delta u = f \chi_{\{u>0\}} \quad \text{in} \quad \Omega. \tag{3.7}$$

(Here, χ_A denotes the characteristic function of a set $A \subset \mathbb{R}^n$.) We will show this in detail later.

3.1.3   The Free Boundary

Let us take a closer look at the obstacle problem (3.3).

One of the most important features of such problem is that it has *two* unknowns: the *solution* u, and the *contact set* $\{u = 0\}$. In other words, there are two regions in Ω: one in which $u = 0$; and one in which $\Delta u = f$.

These regions are characterized by the minimization problem (3.4). Moreover, if we denote

$$\Gamma := \partial\{u > 0\} \cap \Omega,$$

then this is called the *free boundary*, see Fig. 3.2.

The obstacle problem is a *free boundary problem*, as it involves an *unknown interface* Γ as part of the problem.

Moreover, if we assume Γ to be smooth, then it is easy to see that the fact that u is a nonnegative supersolution must imply $\nabla u = 0$ on Γ, that is, we will have that $u \geq 0$ solves

$$\begin{cases} \Delta u = f \text{ in } \{u > 0\} \\ \quad u = 0 \text{ on } \Gamma \\ \nabla u = 0 \text{ on } \Gamma. \end{cases} \tag{3.8}$$

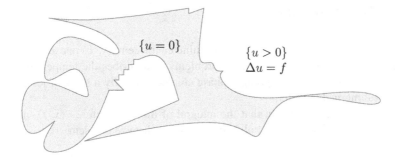

$\{u = 0\}$

$\{u > 0\}$
$\Delta u = f$

Fig. 3.2 The free boundary could, a priori, be very irregular

This is an alternative way to write the Euler–Lagrange equation of the problem. In this way, the interface Γ appears clearly, and we see that we have *both* Dirichlet *and* Neumann conditions on Γ.

This would usually be an over-determined problem (too many boundary conditions on Γ), but since Γ is also free, it turns out that the problem has a unique solution (where Γ is part of the solution, of course).

3.2 Some Motivations and Applications

Let us briefly comment on some of the main motivations and applications in the study of the obstacle problem. We refer to the books [14, 16, 20, 22, 28, 29], for more details and further applications of obstacle-type problems.

3.2.1 Fluid Filtration

The so-called Dam problem aims to describe the filtration of water inside a porous dam. One considers a dam separating two reservoirs of water at different heights, made of a porous medium (permeable to water). Then there is some transfer of water across the dam, and the interior of the dam has a wet part, where water flows, and a dry part. In this setting, an integral of the pressure (with respect to the height of the column of water at each point) solves the obstacle problem, and the free boundary corresponds precisely to the interphase separating the wet and dry parts of the dam.

3.2.2   Phase Transitions

The Stefan problem, dating back to the nineteenth century, is one of the most classical and important free boundary problems. It describes the temperature of a homogeneous medium undergoing a phase change, typically a body of ice at zero degrees submerged in water.

In this context, it turns out that the integral of the temperature $\theta(x, t)$, namely $u(x, t) := \int_0^t \theta$, solves the parabolic version of the obstacle problem,

$$u_t - \Delta u = \chi_{\{u>0\}} \quad \text{in} \quad \Omega \times (0, T) \subset \mathbb{R}^3 \times \mathbb{R},$$
$$\partial_t u \geq 0,$$
$$u \geq 0.$$

The moving interphase separating the solid and liquid is exactly the free boundary $\partial\{u > 0\}$.

3.2.3   Hele-Shaw Flow

This 2D model, dating back to 1898, describes a fluid flow between two flat parallel plates separated by a very thin gap. Various problems in fluid mechanics can be approximated to Hele-Shaw flows, and that is why understanding these flows is important.

A Hele-Shaw cell is an experimental device in which a viscous fluid is sandwiched in a narrow gap between two parallel plates. In certain regions, the gap is filled with fluid while in others the gap is filled with air. When liquid is injected inside the device though some sinks (e.g. though a small hole on the top plate) the region filled with liquid grows. In this context, an integral of the pressure solves, for each fixed time t, the obstacle problem. In a similar way to the Dam problem, the free boundary corresponds to the interface between the fluid and the air regions.

3.2.4   Optimal Stopping, Finance

In probability and finance, the obstacle problem appears when considering optimal stopping problems for stochastic processes.

Indeed, consider a random walk (Brownian motion) inside a domain $\Omega \subset \mathbb{R}^n$, and a payoff function φ defined on the same domain. We can stop the random walk at any moment, and we get the payoff at that position. We want to maximize the expected payoff (by choosing appropriately the stopping strategy). Then, it turns out that the highest expected payoff $v(x)$ starting at a given position x satisfies the obstacle problem (3.2), where the contact set $\{v = \varphi\}$ is the region where we should immediately stop the random walk and get the payoff, while $\{v > \varphi\}$ is the region where we should wait.

3.2.5 Interacting Particle Systems

Large systems of interacting particles arise in physical, biological, or material sciences.

In some models the particles attract each other when they are far, and experience a repulsive force when they are close. In other related models in statistical mechanics, the particles (e.g. electrons) repel with a Coulomb force and one wants to understand their behavior in presence of some external field that confines them.

In this kind of models, a natural and interesting question is to determine the "equilibrium configurations". For instance, in Coulomb systems the charges accumulate in some region with a well defined boundary. Interestingly, these problems are equivalent to the obstacle problem—namely, the electric potential $u = u(x)$ generated by the charges solves such problem—and the contact set $\{u = 0\}$ corresponds to the region in which the particles concentrate.

3.3 Basic Properties of Solutions

We proceed now to study the basic properties of solutions $u \geq 0$ to the obstacle problem (3.4) or (3.6).

3.3.1 Existence of Solutions

Since problem (3.4) is a minimization of a convex functional, the existence and uniqueness of solutions is standard.

Proposition 1 (Existence and uniqueness) *Let $\Omega \subset \mathbb{R}^n$ be any bounded Lipschitz domain, and let $g : \partial\Omega \to \mathbb{R}$ be such that*

$$C = \{u \in H^1(\Omega) : u \geq 0 \text{ in } \Omega, \ u|_{\partial\Omega} = g\} \neq \varnothing.$$

Then, for any $f \in L^2(\Omega)$ there exists a unique minimizer of

$$\frac{1}{2} \int_\Omega |\nabla u|^2 dx + \int_\Omega fu$$

among all functions u satisfying $u \geq 0$ in Ω and $u|_{\partial\Omega} = g$.

The proof is left as an exercise to the reader.

Furthermore, we have the following equivalence. (Recall that we denote $u^+ = \max\{u, 0\}$, and $u^- = \max\{-u, 0\}$, so that $u = u^+ - u^-$.)

Proposition 2 *Let $\Omega \subset \mathbb{R}^n$ be any bounded Lipschitz domain, and let $g : \partial\Omega \to \mathbb{R}$ be such that*

$$C = \left\{u \in H^1(\Omega) : u \geq 0 \text{ in } \Omega, \ u|_{\partial\Omega} = g\right\} \neq \varnothing.$$

Then, the following are equivalent.

(i) *u minimizes $\frac{1}{2}\int_\Omega |\nabla u|^2 + \int_\Omega fu$ among all functions satisfying $u \geq 0$ in Ω and $u|_{\partial\Omega} = g$.*

(ii) *u minimizes $\frac{1}{2}\int_\Omega |\nabla u|^2 + \int_\Omega fu^+$ among all functions satisfying $u|_{\partial\Omega} = g$.*

Proof The two functionals coincide whenever $u \geq 0$. Thus, the only key point is to prove that the minimizer in (ii) must be nonnegative, i.e., $u = u^+$. (Notice that since $C \neq \varnothing$ then $g \geq 0$ on $\partial\Omega$.) To show this, recall that the positive part of any H^1 function is still in H^1, and moreover $|\nabla u|^2 = |\nabla u^+|^2 + |\nabla u^-|^2$. Thus, we have that

$$\frac{1}{2}\int_\Omega |\nabla u^+|^2 + \int_\Omega fu^+ \leq \frac{1}{2}\int_\Omega |\nabla u|^2 + \int_\Omega fu^+,$$

with strict inequality unless $u = u^+$. This means that any minimizer u of the functional in (ii) must be nonnegative, and thus we are done.

We next notice that any minimizer of (3.4) is actually a solution to (3.9) below. Recall that we always assuming that obstacles are as smooth as necessary, $\varphi \in C^\infty(\Omega)$, and therefore we assume here that $f \in C^\infty(\Omega)$ as well.

Proposition 3 *Let $\Omega \subset \mathbb{R}^n$ be any bounded Lipschitz domain, $f \in C^\infty(\Omega)$, and u be any minimizer of (3.4) subject to the boundary conditions $u|_{\partial\Omega} = g$.*

Then, u solves

$$\Delta u = f\chi_{\{u>0\}} \quad \text{in} \quad \Omega \tag{3.9}$$

in the weak sense.

Proof This is left as an exercise to the reader.

As a consequence, we find:

Corollary 1 *Let $\Omega \subset \mathbb{R}^n$ be any bounded Lipschitz domain, $f \in C^\infty(\Omega)$, and u be any minimizer of (3.4) subject to the boundary conditions $u|_{\partial\Omega} = g$.*

Then, u is $C^{1,\alpha}$ inside Ω, for every $\alpha \in (0, 1)$.

Proof By Proposition 3, u solves

$$\Delta u = f \chi_{\{u>0\}} \quad \text{in} \quad \Omega.$$

Since $f \chi_{\{u>0\}} \in L^\infty(\Omega)$, then by standard regularity estimates we deduce that $u \in C^{1,1-\varepsilon}$ for every $\varepsilon > 0$.

3.3.2 Optimal Regularity of Solutions

Thanks to the previous results, we know that any minimizer of (3.4) solves (3.9) and is $C^{1,\alpha}$. From now on, we will localize the problem and study it in a ball:

$$\begin{aligned} u &\geq 0 && \text{in } B_1 \\ \Delta u &= f \chi_{\{u>0\}} && \text{in } B_1. \end{aligned} \tag{3.10}$$

Our next goal is to answer the following question:

Question : *What is the optimal regularity of solutions?*

First, a few important considerations. Notice that in the set $\{u > 0\}$ we have $\Delta u = f$, while in the interior of $\{u = 0\}$ we have $\Delta u = 0$ (since $u \equiv 0$ there).

Thus, since f is in general not zero, then Δu is *discontinuous* across the free boundary $\partial\{u > 0\}$ in general. In particular, $u \notin C^2$.

We will now prove that any minimizer of (3.4) is actually $C^{1,1}$, which gives the:

Answer : $u \in C^{1,1}$ *(second derivatives are bounded but not continuous)*

The precise statement and proof are given next.

Theorem 1 (Optimal regularity) *Let $f \in C^\infty(B_1)$, and u be any solution to (3.10). Then, u is $C^{1,1}$ inside $B_{1/2}$, with the estimate*

$$\|u\|_{C^{1,1}(B_{1/2})} \leq C\big(\|u\|_{L^\infty(B_1)} + \|f\|_{\mathrm{Lip}(B_1)}\big).$$

The constant C depends only on n.

To prove this, the main step is the following.

Lemma 1 *Let u be any solution to (3.10). Let $x_\circ \in \overline{B_{1/2}}$ be any point on $\{u = 0\}$. Then, for any $r \in (0, \frac{1}{4})$ we have*

$$0 \leq \sup_{B_r(x_\circ)} u \leq Cr^2,$$

with C depending only on n and $\|f\|_{L^\infty(B_1)}$.

Proof We have that $\Delta u = f \chi_{\{u>0\}}$ in B_1, with $f \chi_{\{u>0\}} \in L^\infty(B_1)$. Thus, since $u \geq 0$, we can use the Harnack inequality for the equation $\Delta u = f \chi_{\{u>0\}}$ in $B_{2r}(x_o)$, to find

$$\sup_{B_r(x_o)} u \leq C \left(\inf_{B_r(x_o)} u + r^2 \| f \chi_{\{u>0\}} \|_{L^\infty(B_{2r}(x_o))} \right).$$

Since $u \geq 0$ and $u(x_o) = 0$, this yields $\sup_{B_r(x_o)} u \leq C \| f \|_{L^\infty(B_1)} r^2$, as wanted.

We have proved that:

At every free boundary point x_o, u grows (at most) quadratically.

As shown next, this easily implies the $C^{1,1}$ regularity.

Proof *(Proof of Theorem 1)* Dividing u by a constant if necessary, we may assume that $\|u\|_{L^\infty(B_1)} + \|f\|_{C^{0,\alpha}(B_1)} \leq 1$, where $\alpha \in (0,1)$ is fixed.

We already know that $u \in C^\infty$ in the set $\{u > 0\}$ (since $\Delta u = f \in C^\infty$), and also inside the set $\{u = 0\}$ (since $u = 0$ there). Moreover, on the interface $\Gamma = \partial\{u > 0\}$ we have proved the quadratic growth $\sup_{B_r(x_o)} u \leq Cr^2$. Let us prove that this yields the $C^{1,1}$ bound we want.

Let $x_1 \in \{u > 0\} \cap B_{1/2}$, and let $x_o \in \Gamma$ be the closest free boundary point. Denote $\rho = |x_1 - x_o|$. Then, we have $\Delta u = f$ in $B_\rho(x_1)$.

By Schauder estimates, we find

$$\|D^2 u\|_{L^\infty(B_{\rho/2}(x_1))} \leq C \left(\frac{1}{\rho^2} \|u\|_{L^\infty(B_\rho(x_1))} + \|f\|_{C^{0,\alpha}(B_1)} \right).$$

But by the growth proved in the previous Lemma, we have $\|u\|_{L^\infty(B_\rho(x_1))} \leq C\rho^2$, which yields

$$\|D^2 u\|_{L^\infty(B_{\rho/2}(x_1))} \leq C.$$

In particular,

$$|D^2 u(x_1)| \leq C.$$

Since we can do this for each $x_1 \in \{u > 0\} \cap B_{1/2}$, it follows that $\|u\|_{C^{1,1}(B_{1/2})} \leq C$, as wanted.

The overall strategy of the proof of optimal regularity is summarized in Fig. 3.3.

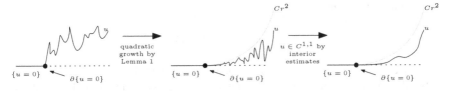

Fig. 3.3 Strategy of the proof of Theorem 1

3.3.3 Nondegeneracy

We now want to prove that, at all free boundary points, u grows *at least* quadratically (we already know *at most* quadratically).

That is, we want

$$0 < cr^2 \leq \sup_{B_r(x_o)} u \leq Cr^2 \tag{3.11}$$

for all free boundary points $x_o \in \partial\{u > 0\}$.

This property is essential in order to study the free boundary later.

For this, we need the following:

Assumption *The right hand side f satisfies*

$$f \geq c_o > 0$$

in the ball B_1.

(Actually, it is common to simply assume $f \equiv 1$, since this is the right hand side that arises naturally in many models.)

Proposition 4 (Nondegeneracy) *Let u be any solution to (3.10). Assume that $f \geq c_o > 0$ in B_1. Then, for every free boundary point $x_o \in \partial\{u > 0\} \cap B_{1/2}$, we have*

$$0 < cr^2 \leq \sup_{B_r(x_o)} u \leq Cr^2 \qquad \text{for all } r \in (0, \tfrac{1}{2}),$$

with a constant $c > 0$ depending only on n and c_o.

Proof Let $x_1 \in \{u > 0\}$ be any point close to x_o (we will then let $x_1 \to x_o$ at the end of the proof).

Consider the function

$$w(x) := u(x) - \frac{c_o}{2n}|x - x_1|^2.$$

Then, in $\{u > 0\}$ we have

$$\Delta w = \Delta u - c_\circ = f - c_\circ \geq 0$$

and hence $-\Delta w \leq 0$ in $\{u > 0\} \cap B_r(x_1)$. Moreover, $w(x_1) > 0$.

By the maximum principle, w attains a positive maximum on $\partial(\{u > 0\} \cap B_r(x_1))$. But on the free boundary $\partial\{u > 0\}$ we clearly have $w < 0$. Therefore, there is a point on $\partial B_r(x_1)$ at which $w > 0$. In other words,

$$0 < \sup_{\partial B_r(x_1)} w = \sup_{\partial B_r(x_1)} u - \frac{c_\circ}{2n} r^2.$$

Letting now $x_1 \to x_\circ$, we find $\sup_{\partial B_r(x_\circ)} u \geq cr^2 > 0$, as desired.

3.3.4   Summary of Basic Properties

Let u be any solution to the obstacle problem

$$
\begin{aligned}
u &\geq 0 && \text{in } B_1, \\
\Delta u &= f \chi_{\{u>0\}} && \text{in } B_1.
\end{aligned}
$$

Then, we have:

- Optimal regularity: $\|u\|_{C^{1,1}(B_{1/2})} \leq C \left(\|u\|_{L^\infty(B_1)} + \|f\|_{\mathrm{Lip}(B_1)} \right)$
- Nondegeneracy: If $f \geq c_\circ > 0$, then

$$0 < cr^2 \leq \sup_{B_r(x_\circ)} u \leq Cr^2 \qquad \text{for all } r \in (0, \tfrac{1}{2})$$

at all free boundary points $x_\circ \in \partial\{u > 0\} \cap B_{1/2}$.

Using these properties, we can now start the study of the free boundary.

3.4   Regularity of Free Boundaries: An Overview

From now on, we consider any solution to

$$
\begin{aligned}
&u \in C^{1,1}(B_1), \\
&u \geq 0 \quad \text{in } B_1, \\
&\Delta u = f \quad \text{in } \{u > 0\},
\end{aligned}
\tag{3.12}
$$

Fig. 3.4 A solution to the
obstacle problem in B_1

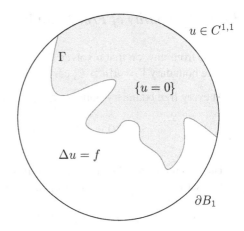

(see Fig. 3.4) with

$$f \geq c_\circ > 0 \qquad \text{and} \qquad f \in C^\infty. \tag{3.13}$$

Notice that on the interface

$$\Gamma = \partial\{u > 0\} \cap B_1$$

we have that

$$u = 0 \quad \text{on } \Gamma,$$

$$\nabla u = 0 \quad \text{on } \Gamma.$$

The central mathematical challenge in the obstacle problem is to

Understand the geometry/regularity of the free boundary Γ.

Notice that, even if we already know the optimal regularity of u (it is $C^{1,1}$), we know nothing about the free boundary Γ. A priori Γ could be a very irregular object, even a fractal set with infinite perimeter.

As we will see, under the natural assumption $f \geq c_\circ > 0$, it turns out that free boundaries are always smooth, possibly outside a certain set of singular points. In fact, in our proofs we will assume for simplicity that $f \equiv 1$ (or constant). We do that in order to avoid x-dependence and the technicalities associated to it, which gives cleaner proofs. In this way, the main ideas behind the regularity of free boundaries are exposed.

3.4.1 Regularity of Free Boundaries: Main Results

Assume from now on that u solves (3.12)–(3.13). Then, the main known results on the free boundary $\Gamma = \partial\{u > 0\}$ can be summarized as follows:

- At every free boundary point $x_o \in \Gamma$, we have

$$0 < cr^2 \le \sup_{B_r(x_o)} u \le Cr^2 \qquad\qquad \forall r \in (0, \tfrac{1}{2})$$

- The free boundary Γ splits into *regular points* and *singular points*.
- The set of *regular points* is an open subset of the free boundary, and Γ is C^∞ near these points.
- *Singular points* are those at which the contact set $\{u = 0\}$ has *density zero*, and these points (if any) are contained in an $(n-1)$-dimensional C^1 manifold.

Summarizing, *the free boundary is smooth, possibly outside a certain set of singular points*. See Fig. 3.5.

So far, we have not even proved that Γ has finite perimeter, or anything at all about Γ. Our goal will be to prove that Γ *is C^∞ near regular points*. This is the main and most important result in the obstacle problem. It was proved by Caffarelli in 1977, and it is one of the major results for which he received the Wolf Prize in 2012 and the Shaw Prize in 2018.

3.4.2 Overview of the Strategy

To prove these regularity results for the free boundary, one considers *blow-ups*. Namely, given any free boundary point x_o for a solution u of (3.12)–(3.13), one takes the rescalings

$$u_r(x) := \frac{u(x_o + rx)}{r^2},$$

with $r > 0$ small. This is like "zooming in" at a free boundary point.

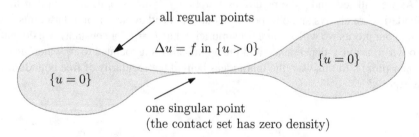

Fig. 3.5 Singular points are those where the contact set has zero density

The factor r^{-2} is chosen so that

$$\|u_r\|_{L^\infty(B_1)} \approx 1$$

as $r \to 0$; recall that $0 < cr^2 \le \sup_{B_r(x_\circ)} u \le Cr^2$.

Then, by $C^{1,1}$ estimates, we will prove that a subsequence of u_r converges to a function u_0 locally uniformly in $\mathbb{R}^n$ as $r \to 0$. Such function u_0 is called a *blow-up of u at $x_\circ$*.

Any blow-up u_0 is a *global* solution to the obstacle problem, with $f \equiv 1$ (or with $f \equiv \mathrm{ctt} > 0$).

Then, the main issue is to *classify blow-ups*: that is, to show that

either	$u_0(x) = \frac{1}{2}(x \cdot e)_+^2$	(this happens at regular points)
or	$u_0(x) = \frac{1}{2}x^T A x$	(this happens at singular points).

Here, $e \in \mathbb{S}^{n-1}$ is a unit vector, and $A \ge 0$ is a positive semi-definite matrix satisfying $\mathrm{tr} A = 1$. Notice that the contact set $\{u_0 = 0\}$ becomes a half-space in case of regular points, while it has zero measure in case of singular points; see Fig. 3.6.

Once this is done, one has to "transfer" the information from the blow-up u_0 to the original solution u. Namely, one shows that, in fact, the free boundary is $C^{1,\alpha}$ near regular points (for some small $\alpha > 0$).

Finally, once we know that the free boundary is $C^{1,\alpha}$, then we will "bootstrap" the regularity to C^∞, by using fine estimates for harmonic functions in $C^{k,\alpha}$ domains.

Classifying blow-ups is not easy. Generally speaking, classifying blow-ups is of similar difficulty to proving regularity estimates.

Thus, how can we classify blow-ups? Do we get any extra information on u_0 that we did not have for u? (Otherwise it seems hopeless!)

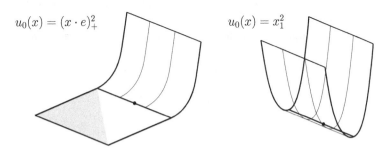

Fig. 3.6 Possible blow-ups of the solution to the obstacle problem at free boundary points

The answer is *yes*: CONVEXITY. We will prove that all blow-ups are always *convex*. This is a huge improvement, since this yields that the contact set $\{u_0 = 0\}$ is also convex. Furthermore, we will show that blow-ups are also *homogeneous*.

So, before the blow-up we had no information on the set $\{u = 0\}$, but after the blow-up we get that $\{u_0 = 0\}$ *is a convex cone*. Thanks to this we will be able to classify blow-ups, and thus to prove the regularity of the free boundary.

The main steps in the proof of the regularity of the free boundary will be the following:

1. $0 < cr^2 \leq \sup_{B_r(x_\circ)} u \leq Cr^2$
2. Blow-ups u_0 are *convex* and *homogeneous*.
3. If the contact set has *positive density* at $x_\circ$, then $u_0(x) = \frac{1}{2}(x \cdot e)_+^2$.
4. Deduce that the free boundary is $C^{1,\alpha}$ near $x_\circ$.
5. Deduce that the free boundary is C^∞ near $x_\circ$.

The proofs that we will present here are a modified version of the original ones due to Caffarelli (see [6]), together with some extra tools due to Weiss (see [36]). We refer to [6, 28], and [36], for different proofs of the classification of blow-ups and/or of the regularity of free boundaries.

3.5   Classification of Blow-Ups

The aim of this section is to classify all possible blow-ups u_0.

3.5.1   Convexity of Blow-Ups

The first important property about blow-ups is that they are convex. More precisely, for the original solution u in B_1, the closer we look to a free boundary point $x_\circ$, the closer is the solution to being convex.

Recall that, for simplicity, from now on we will assume that $f \equiv 1$ in B_1. This is only to avoid x-dependence in the equation.

Therefore, from now on we consider a solution u satisfying (see Fig. 3.7):

$$
\begin{aligned}
&u \in C^{1,1}(B_1) \\
&u \geq 0 \quad \text{in } B_1 \\
&\Delta u = 1 \quad \text{in } \{u > 0\} \\
&0 \text{ is a free boundary point.}
\end{aligned}
\tag{3.14}
$$

We will prove all the results around the origin (without loss of generality).

The convexity of blow-ups is given by the following.

Fig. 3.7 A solution u to the obstacle problem with $f \equiv 1$

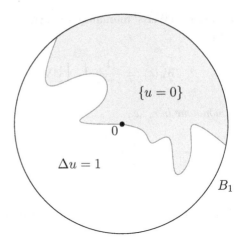

Theorem 2 *Let $u_0 \in C^{1,1}$ be any global solution to*

$$u_0 \geq 0 \quad \text{in } \mathbb{R}^n$$
$$\Delta u_0 = 1 \quad \text{in } \{u_0 > 0\}$$
$$0 \text{ is a free boundary point.}$$

Then, u_0 is convex.

We skip the proof of such result, and we refer to [6, 16] or [28] for more details.

3.5.2   Homogeneity of Blow-Ups

We will now prove that blow-ups are homogeneous. This is not essential in the proof of the regularity of the free boundary (see [6]), but it actually simplifies it. We will show that, for the original solution u in B_1, the closer we look at a free boundary point $x_\circ$, the closer is the solution to being homogeneous.

Proposition 5 (Homogeneity of blow-ups) *Let u be any solution to (3.14). Then, any blow-up of u at 0 is homogeneous of degree 2.*

It is important to remark that not all global solutions to the obstacle problem in $\mathbb{R}^n$ are homogeneous. There exist global solutions u_0 that are convex, $C^{1,1}$, and whose contact set $\{u_0 = 0\}$ is an ellipsoid, for example. However, thanks to the previous result, we find that such non-homogeneous solutions cannot appear as blow-ups, i.e., that all blow-ups must be homogeneous.

To prove this, we will need the following monotonicity formula due to Weiss.

Theorem 3 (Weiss' monotonicity formula) *Let u be any solution to (3.14). Then, the quantity*

$$W_u(r) := \frac{1}{r^{n+2}} \int_{B_r} \left\{ \tfrac{1}{2} |\nabla u|^2 + u \right\} - \frac{1}{r^{n+3}} \int_{\partial B_r} u^2 \tag{3.15}$$

is monotone in r, i.e.,

$$\frac{d}{dr} W_u(r) := \frac{1}{r^{n+3}} \int_{\partial B_r} (x \cdot \nabla u - 2u)^2 dx \geq 0$$

for $r \in (0, 1)$.

Proof Let $u_r(x) = r^{-2} u(rx)$, and observe that

$$W_u(r) = \int_{B_1} \left\{ \tfrac{1}{2} |\nabla u_r|^2 + u_r \right\} - \int_{\partial B_1} u_r^2.$$

Using this, together with

$$\frac{d}{dr} u_r = \frac{1}{r} \{ x \cdot \nabla u_r - 2u_r \} \tag{3.16}$$

and

$$\frac{d}{dr} (\nabla u_r) = \nabla \frac{d}{dr} u_r,$$

we find

$$\frac{d}{dr} W_u(r) = \int_{B_1} \left\{ \nabla u_r \cdot \nabla \frac{d}{dr} u_r + \frac{d}{dr} u_r \right\} - 2 \int_{\partial B_1} u_r \frac{d}{dr} u_r.$$

Now, integrating by parts we get

$$\int_{B_1} \nabla u_r \cdot \nabla \frac{d}{dr} u_r = - \int_{B_1} \Delta u_r \frac{d}{dr} u_r + \int_{\partial B_1} \partial_\nu (u_r) \frac{d}{dr} u_r.$$

Since $\Delta u_r = 1$ in $\{u_r > 0\}$ and $\frac{d}{dr} u_r = 0$ in $\{u_r = 0\}$, then

$$\int_{B_1} \nabla u_r \cdot \nabla \frac{d}{dr} u_r = - \int_{B_1} \frac{d}{dr} u_r + \int_{\partial B_1} \partial_\nu (u_r) \frac{d}{dr} u_r.$$

Thus, we deduce

$$\frac{d}{dr} W_u(r) = \int_{\partial B_1} \partial_\nu (u_r) \frac{d}{dr} u_r - 2 \int_{\partial B_1} u_r \frac{d}{dr} u_r.$$

Using that on ∂B_1 we have $\partial_\nu = x \cdot \nabla$, combined with (3.16), yields

$$\frac{d}{dr} W_u(r) = \frac{1}{r} \int_{\partial B_1} (x \cdot \nabla u_r - 2u_r)^2,$$

which gives the desired result.

We now give the:

Proof (*Proof of Proposition 5*) Let $u_r(x) = r^{-2} u(rx)$, and notice that we have the scaling property

$$W_{u_r}(\rho) = W_u(\rho r),$$

for any $r, \rho > 0$.

If u_0 is any blow-up of u at 0 then there is a sequence $r_j \to 0$ satisfying $u_{r_j} \to u_0$ in $C^1_{\text{loc}}(\mathbb{R}^n)$. Thus, for any $\rho > 0$ we have

$$W_{u_0}(\rho) = \lim_{r_j \to 0} W_{u_{r_j}}(\rho) = \lim_{r_j \to 0} W_u(\rho r_j) = W_u(0^+).$$

Notice that the limit $W_u(0^+) := \lim_{r \to 0} W_u(r)$ exists, by monotonicity of W.

Hence, the function $W_{u_0}(\rho)$ is *constant* in ρ. However, by Theorem 3 this yields that $x \cdot \nabla u_0 - 2u_0 \equiv 0$ in $\mathbb{R}^n$, and therefore u_0 is homogeneous of degree 2.

Remark 1 Here, we used that a C^1 function u_0 is 2-homogeneous (i.e. $u_0(\lambda x) = \lambda^2 u_0(x)$ for all $\lambda \in \mathbb{R}_+$) if and only if $x \cdot \nabla u_0 \equiv 2u_0$. This is because $\partial_\lambda |_{\lambda=1} \{\lambda^{-2} u(\lambda x)\} = x \cdot \nabla u_0 - 2u_0$.

3.5.3 Classification of Blow-Ups

We next want to classify all possible blow-ups for solutions to the obstacle problem (3.14). First, we will prove the following.

Proposition 6 *Let u be any solution to (3.14), and let*

$$u_r(x) := \frac{u(rx)}{r^2}.$$

Then, for any sequence $r_k \to 0$ there is a subsequence $r_{k_j} \to 0$ such that

$$u_{r_{k_j}} \longrightarrow u_0 \quad \text{in } C^1_{\text{loc}}(\mathbb{R}^n)$$

as $k_j \to \infty$, for some function u_0 satisfying

$$\begin{cases} u_0 \in C_{\text{loc}}^{1,1}(\mathbb{R}^n) \\ u_0 \geq 0 \quad \text{in } B_1 \\ \Delta u_0 = 1 \quad \text{in } \{u_0 > 0\} \\ 0 \text{ is a free boundary point} \\ u_0 \text{ is convex} \\ u_0 \text{ is homogeneous of degree 2.} \end{cases} \tag{3.17}$$

Proof By $C^{1,1}$ regularity of u, and by nondegeneracy, we have that

$$\frac{1}{C} \leq \sup_{B_1} u_r \leq C$$

for some $C > 0$. Moreover, again by $C^{1,1}$ regularity of u, we have

$$\|D^2 u_r\|_{L^\infty(B_{1/r})} \leq C.$$

Since the sequence $\{u_{r_k}\}$, for $r_k \to 0$, is uniformly bounded in $C^{1,1}(K)$ for each compact set $K \subset \mathbb{R}^n$, then there is a subsequence $r_{k_j} \to 0$ such that

$$u_{r_{k_j}} \longrightarrow u_0 \quad \text{in } C_{\text{loc}}^1(\mathbb{R}^n)$$

for some $u_0 \in C^{1,1}(K)$. Moreover, such function u_0 satisfies $\|D^2 u_0\|_{L^\infty(K)} \leq C$, with C independent of K, and clearly $u_0 \geq 0$ in K.

The fact that $\Delta u_0 = 1$ in $\{u_0 > 0\} \cap K$ can be checked as follows. For any smooth function $\eta \in C_c^\infty(\{u > 0\} \cap K)$ we will have that, for k_j large enough, $u_{r_{k_j}} > 0$ in the support of η, and thus

$$\int_{\mathbb{R}^n} \nabla u_{r_{k_j}} \cdot \nabla \eta \, dx = \int_{\mathbb{R}^n} \eta \, dx.$$

Since $u_{r_{k_j}} \to u_0$ in $C^1(K)$ then we can take the limit $k_j \to \infty$ to get

$$\int_{\mathbb{R}^n} \nabla u_0 \cdot \nabla \eta \, dx = \int_{\mathbb{R}^n} \eta \, dx.$$

Since this can be done for any $\eta \in C_c^\infty(\{u > 0\} \cap K)$, and for every $K \subset \mathbb{R}^n$, it follows that $\Delta u_0 = 1$ in $\{u_0 > 0\}$.

The fact that 0 is a free boundary point for u_0 follows simply by taking limits to $u_{r_{k_j}}(0) = 0$ and $\|u_{r_{k_j}}\|_{L^\infty(B_\rho)} \approx \rho^2$ for all $\rho \in (0, 1)$. Finally, the convexity and homogeneity of u_0 follow from Theorem 2 and Proposition 5.

Our next goal is to prove the following.

Theorem 4 (Classification of blow-ups) *Let u be any solution to (3.14), and let u_0 be any blow-up of u at 0. Then,*

(a) *either*

$$u_0(x) = \frac{1}{2}(x \cdot e)_+^2$$

for some $e \in \mathbb{S}^{n-1}$.

(b) *or*

$$u_0(x) = \frac{1}{2}x^T A x$$

for some matrix $A \geq 0$ with $\operatorname{tr} A = 1$.

It is important to remark here that, a priori, different subsequences could lead to different blow-ups u_0.

In order to establish Theorem 4, we will need the following.

Lemma 2 *Let $\Sigma \subset \mathbb{R}^n$ be any closed convex cone with nonempty interior, and with vertex at the origin. Let $w \in C(\mathbb{R}^n)$ be a function satisfying $\Delta w = 0$ in Σ^c, $w > 0$ in Σ^c, and $w = 0$ in Σ.*

Assume in addition that w is homogeneous of degree 1. Then, Σ must be a half-space.

Proof By convexity of Σ, there exists a half-space $H = \{x \cdot e > 0\}$, with $e \in \mathbb{S}^{n-1}$, such that $H \subset \Sigma^c$.

Let $v(x) = (x \cdot e)_+$, which is harmonic and positive in H, and vanishes in H^c. By Hopf Lemma, we have that $w \geq c_o d_\Sigma$ in $\Sigma^c \cap B_1$, where $d_\Sigma(x) = \operatorname{dist}(x, \Sigma)$ and c_o is a small positive constant. In particular, since both w and d_Σ are homogeneous of degree 1, we deduce that $w \geq c_o d_\Sigma$ in all of Σ^c. Notice that, in order to apply Hopf Lemma, we used that—by convexity of Σ—the domain Σ^c satisfies the interior ball condition.

Thus, since $d_\Sigma \geq d_H = v$, we deduce that $w \geq c_o v$, for some $c_o > 0$. The idea is now to consider the functions w and cv, and let $c > 0$ increase until the two functions touch at one point, which will give us a contradiction (recall that two harmonic functions cannot touch at an interior point). To do this rigorously, define

$$c_* := \sup\{c > 0 : w \geq cv \quad \text{in} \quad \Sigma^c\}.$$

Notice that $c_* \geq c_o > 0$. Then, we consider the function $w - c_* v \geq 0$. Assume that $w - c_* v$ is not identically zero. Such function is harmonic in H and hence, by the strict maximum principle, $w - c_* v > 0$ in H. Then, using Hopf Lemma in H we deduce that $w - c_* v \geq c_o d_H = c_o v$, since v is exactly the distance to H^c. But then we get that $w - (c_* + c_o)v \geq 0$, a contradiction with the definition of c_*.

Therefore, it must be $w - c_*v \equiv 0$. This means that w is a multiple of v, and therefore $\Sigma = H$, a half-space.

Remark 2 (Alternative proof) An alternative way to argue in the previous lemma could be the following. Any function w which is harmonic in a cone Σ^c and homogeneous of degree α can be written as a function on the sphere, satisfying $\Delta_{\mathbb{S}^{n-1}} w = \mu w$ on $\mathbb{S}^{n-1} \cap \Sigma^c$ with $\mu = \alpha(n + \alpha - 2)$—in our case $\alpha = 1$. (Here, $\Delta_{\mathbb{S}^{n-1}}$ denotes the spherical Laplacian, i.e. the Laplace–Beltrami operator on $\mathbb{S}^{n-1}$.) In other words, *homogeneous harmonic functions solve an eigenvalue problem on the sphere.*

Using this, we notice that $w > 0$ in Σ^c and $w = 0$ in Σ imply that w is the *first* eigenfunction of $\mathbb{S}^{n-1} \cap \Sigma^c$, and that the first eigenvalue is $\mu = n - 2$. But, on the other hand, the same happens for the domain $H = \{x \cdot e > 0\}$, since $v(x) = (x \cdot e)_+$ is a positive harmonic function in H. This means that both domains $\mathbb{S}^{n-1} \cap \Sigma^c$ and $\mathbb{S}^{n-1} \cap H$ have the same first eigenvalue μ. But then, by strict monotonicity of the first eigenvalue with respect to domain inclusions, we deduce that $H \subset \Sigma^c \implies H = \Sigma^c$, as desired.

We will also need the following.

Lemma 3 *Assume that $\Delta u = 1$ in $\mathbb{R}^n \setminus \partial H$, where $\partial H = \{x_1 = 0\}$ is a hyperplane. If $u \in C^1(\mathbb{R}^n)$, then $\Delta u = 1$ in $\mathbb{R}^n$.*

Proof For any ball $B_R \subset \mathbb{R}^n$, we consider the solution to $\Delta w = 1$ in B_R, $w = u$ on ∂B_R, and define $v = u - w$. Then, we have $\Delta v = 0$ in $B_R \setminus \partial H$, and $v = 0$ on ∂B_R. We want to show that u coincides with w, that is, $v \equiv 0$ in B_R.

For this, notice that since v is bounded then for $\kappa > 0$ large enough we have

$$v(x) \le \kappa(2R - |x_1|) \quad \text{in} \quad B_R,$$

where $2R - |x_1|$ is positive in B_R and harmonic in $B_R \setminus \{x_1 = 0\}$. Thus, we may consider $\kappa^* := \inf\{\kappa \ge 0 : v(x) \le \kappa(2R - |x_1|) \text{ in } B_R\}$. Assume $\kappa^* > 0$. Since v and $2R - |x_1|$ are continuous in B_R, and $v = 0$ on ∂B_R, then we must have a point $p \in B_R$ at which $v(p) = \kappa^*(2R - |p_1|)$. Moreover, since v is C^1, and the function $2R - |x_1|$ has a wedge on $H = \{x_1 = 0\}$, then we must have $p \in B_R \setminus H$. However, this is not possible, as two harmonic functions cannot touch tangentially at an interior point p. This means that $\kappa^* = 0$, and hence $v \le 0$ in B_R. Repeating the same argument with $-v$ instead of v, we deduce that $v \equiv 0$ in B_R, and thus the lemma is proved.

Finally, we will use the following basic property of convex functions, whose proof is left as an exercise to the reader.

Lemma 4 *Let $u : \mathbb{R}^n \to \mathbb{R}$ be a convex function such that the set $\{u = 0\}$ contains the straight line $\{te' : t \in \mathbb{R}\}$, $e' \in \mathbb{S}^{n-1}$. Then, $u(x + te') = u(x)$ for all $x \in \mathbb{R}^n$ and all $t \in \mathbb{R}$.*

Proof After a rotation, we may assume $e' = e_n$. Then, writing $x = (x', x_n) \in \mathbb{R}^{n-1} \times \mathbb{R}$, we have that $u(0, x_n) = 0$ for all $x_n \in \mathbb{R}$, and we want to prove that $u(x', x_n) = u(x', 0)$ for all $x' \in \mathbb{R}^{n-1}$ and all $x_n \in \mathbb{R}$.

Now, by convexity, given x' and x_n, for every $\varepsilon > 0$ and $M \in \mathbb{R}$ we have

$$(1 - \varepsilon)u(x', x_n) + \varepsilon u(0, x_n + M) \geq u((1 - \varepsilon)x', x_n + \varepsilon M).$$

Since $u(0, x_n + M) = 0$, then choosing $M = \lambda/\varepsilon$ and letting $\varepsilon \to 0$ we deduce that

$$u(x', x_n) \geq u(x', x_n + \lambda).$$

Finally, since this can be done for any $\lambda \in \mathbb{R}$ and $x_n \in \mathbb{R}$, the result follows.

We finally establish the classification of blow-ups at regular points.

Proof (*Proof of Theorem* 4) Let u_0 be any blow-up of u at 0. We already proved that u_0 is convex and homogeneous of degree 2. We divide the proof into two cases.

Case 1. Assume that $\{u_0 = 0\}$ has nonempty interior. Then, we have $\{u_0 = 0\} = \Sigma$, a closed convex cone with nonempty interior.

For any direction $\tau \in \mathbb{S}^{n-1}$ such that $-\tau \in \overset{\circ}{\Sigma}$, we claim that $\partial_\tau u_0 \geq 0$ in $\mathbb{R}^n$. Indeed, for every $x \in \mathbb{R}^n$ we have that $u_0(x + \tau t)$ is zero for $t \ll -1$, and therefore by convexity of u_0 we get that $\partial_t u_0(x + \tau t)$ is monotone nondecreasing in t, and zero for $t \ll -1$. This means that $\partial_t u_0 \geq 0$, and thus $\partial_\tau u_0 \geq 0$ in $\mathbb{R}^n$, as claimed.

Now, for any such τ, we define $w := \partial_\tau u_0 \geq 0$. Notice that, at least for some $\tau \in \mathbb{S}^{n-1}$ with $-\tau \in \overset{\circ}{\Sigma}$, the function w is not identically zero. Moreover, since it is harmonic in Σ^c—recall that $\Delta u_0 = 1$ in Σ^c—then $w > 0$ in Σ^c.

But then, since w is homogeneous of degree 1, we can apply Lemma 2 to deduce that we must necessarily have that Σ is a half-space.

By convexity of u_0 and Lemma 4, this means that u_0 is a 1D function, i.e., $u_0(x) = U(x \cdot e)$ for some $U : \mathbb{R} \to \mathbb{R}$ and some $e \in \mathbb{S}^{n-1}$. Thus, we have that $U \in C^{1,1}$ solves $U''(t) = 1$ for $t > 0$, with $U(t) = 0$ for $t \leq 0$. We deduce that $U(t) = \frac{1}{2}t_+^2$, and therefore $u_0(x) = \frac{1}{2}(x \cdot e)_+^2$.

Case 2. Assume now that $\{u_0 = 0\}$ has empty interior. Then, by convexity, $\{u_0 = 0\}$ is contained in a hyperplane ∂H. Hence, $\Delta u_0 = 1$ in $\mathbb{R}^n \setminus \partial H$, with ∂H being a hyperplane, and $u_0 \in C^{1,1}$. It follows from Lemma 3 that $\Delta u_0 = 1$ in all of $\mathbb{R}^n$. But then all second derivatives of u_0 are harmonic and globally bounded in $\mathbb{R}^n$, so they must be constant. Hence, u_0 is a quadratic polynomial. Finally, since $u_0(0) = 0$, $\nabla u_0(0) = 0$, and $u_0 \geq 0$, then it must be $u_0(x) = \frac{1}{2}x^T A x$ for some $A \geq 0$, and since $\Delta u_0 = 1$ then $\operatorname{tr} A = 1$.

3.6   Regularity of the Free Boundary

The aim of this section is to prove Theorem 6 below, i.e., that if u is any solution to (3.14) satisfying

$$\limsup_{r \to 0} \frac{|\{u = 0\} \cap B_r|}{|B_r|} > 0 \tag{3.18}$$

(i.e., the contact set has positive density at the origin), then the free boundary $\partial\{u > 0\}$ is C^∞ in a neighborhood of the origin.

For this, we will use the classification of blow-ups established in the previous section.

3.6.1 $C^{1,\alpha}$ Regularity of the Free Boundary

The first step here is to transfer the local information on u given by (3.18) into a blow-up u_0. More precisely, we next show that

$$(3.18) \quad \Longrightarrow \quad \begin{array}{c} \text{The contact set of a blow-up } u_0 \\ \text{has nonempty interior.} \end{array}$$

Lemma 5 *Let u be any solution to (3.14), and assume that (3.18) holds. Then, there is at least one blow-up u_0 of u at 0 such that the contact set $\{u_0 = 0\}$ has nonempty interior.*

Proof Let $r_k \to 0$ be a sequence along which

$$\lim_{r_k \to 0} \frac{|\{u = 0\} \cap B_{r_k}|}{|B_{r_k}|} \geq \theta > 0.$$

Such sequence exists (with $\theta > 0$ small enough) by assumption (3.18).

Recall that, thanks to Proposition 6, there exists a subsequence $r_{k_j} \downarrow 0$ along which $u_{r_{k_j}} \to u_0$ uniformly on compact sets of $\mathbb{R}^n$, where $u_r(x) = r^{-2}u(rx)$ and u_0 is convex.

Assume by contradiction that $\{u_0 = 0\}$ has empty interior. Then, by convexity, we have that $\{u_0 = 0\}$ is contained in a hyperplane, say $\{u_0 = 0\} \subset \{x_1 = 0\}$.

Since $u_0 > 0$ in $\{x_1 \neq 0\}$ and u_0 is continuous, we have that for each $\delta > 0$

$$u_0 \geq \varepsilon > 0 \quad \text{in } \{|x_1| > \delta\} \cap B_1$$

for some $\varepsilon > 0$.

Therefore, by uniform convergence of $u_{r_{k_j}}$ to u_0 in B_1, there is $r_{k_j} > 0$ small enough such that

$$u_{r_{k_j}} \geq \frac{\varepsilon}{2} > 0 \quad \text{in } \{|x_1| > \delta\} \cap B_1.$$

In particular, the contact set of $u_{r_{k_j}}$ is contained in $\{|x_1| \leq \delta\} \cap B_1$, so

$$\frac{\left|\{u_{r_{k_j}} = 0\} \cap B_1\right|}{|B_1|} \leq \frac{\left|\{|x_1| \leq \delta\} \cap B_1\right|}{|B_1|} \leq C\delta.$$

Rescaling back to u, we find

$$\frac{\left|\{u = 0\} \cap B_{r_{k_j}}\right|}{|B_{r_{k_j}}|} = \frac{\left|\{u_{r_{k_j}} = 0\} \cap B_1\right|}{|B_1|} < C\delta.$$

Since we can do this for every $\delta > 0$, we find that $\lim_{r_{k_j} \to 0} \frac{|\{u=0\} \cap B_{r_{k_j}}|}{|B_{r_{k_j}}|} = 0$, a contradiction. Thus, the lemma is proved.

Combining the previous lemma with the classification of blow-ups from the previous section, we deduce:

Corollary 2 *Let u be any solution to (3.14), and assume that (3.18) holds. Then, there is at least one blow-up of u at 0 of the form*

$$u_0(x) = \frac{1}{2}(x \cdot e)_+^2, \qquad e \in \mathbb{S}^{n-1}. \qquad (3.19)$$

Proof The result follows from Lemma 5 and Theorem 4.

We now want to use this information to show that the free boundary must be smooth in a neighborhood of 0. For this, we start with the following.

Proposition 7 *Let u be any solution to (3.14), and assume that (3.18) holds. Fix any $\varepsilon > 0$. Then, there exist $e \in \mathbb{S}^{n-1}$ and $r_o > 0$ such that*

$$\left|u_{r_o}(x) - \frac{1}{2}(x \cdot e)_+^2\right| \leq \varepsilon \qquad \text{in } B_1,$$

and

$$\left|\partial_\tau u_{r_o}(x) - (x \cdot e)_+(\tau \cdot e)\right| \leq \varepsilon \qquad \text{in } B_1$$

for all $\tau \in \mathbb{S}^{n-1}$.

Proof By Corollary 2 and Proposition 6, we know that there is a subsequence $r_j \to$ 0 for which $u_{r_j} \to \frac{1}{2}(x \cdot e)_+^2$ in $C^1_{\mathrm{loc}}(\mathbb{R}^n)$, for some $e \in \mathbb{S}^{n-1}$. In particular, for every $\tau \in \mathbb{S}^{n-1}$ we have $u_{r_j} \to \frac{1}{2}(x \cdot e)_+^2$ and $\partial_\tau u_{r_j} \to \partial_\tau \left[\frac{1}{2}(x \cdot e)_+^2\right]$ uniformly in B_1.

This means that, given $\varepsilon > 0$, there exists j_o such that

$$\left|u_{r_{j_o}}(x) - \tfrac{1}{2}(x \cdot e)_+^2\right| \leq \varepsilon \qquad \text{in} \quad B_1,$$

and

$$\left|\partial_\tau u_{r_{j_o}}(x) - \partial_\tau\left[\tfrac{1}{2}(x \cdot e)_+^2\right]\right| \leq \varepsilon \qquad \text{in} \quad B_1.$$

Since $\partial_\tau\left[\frac{1}{2}(x \cdot e)_+^2\right] = (x \cdot e)_+(\tau \cdot e)$, then the proposition is proved.

Now, notice that if $(\tau \cdot e) > 0$, then the derivatives $\partial_\tau u_0 = (x \cdot e)_+(\tau \cdot e)$ are *nonnegative*, and strictly positive in $\{x \cdot e > 0\}$ (see Fig. 3.8).

We want to transfer this information to u_{r_o}, and prove that $\partial_\tau u_{r_o} \geq 0$ in B_1 for all $\tau \in \mathbb{S}^{n-1}$ satisfying $\tau \cdot e \geq \frac{1}{2}$. For this, we need a lemma.

Lemma 6 *Let u be any solution to (3.14), and consider $u_{r_o}(x) = r_o^{-2} u(r_o x)$ and $\Omega = \{u_{r_o} > 0\}$.*

Assume that a function $w \in C(B_1)$ satisfies:

(a) *w is bounded and harmonic in $\Omega \cap B_1$.*
(b) *$w = 0$ on $\partial\Omega \cap B_1$.*
(c) *Denoting $N_\delta := \{x \in B_1 : \mathrm{dist}(x, \partial\Omega) < \delta\}$, we have*

$$w \geq -c_1 \quad \text{in} \quad N_\delta \qquad \text{and} \qquad w \geq C_2 > 0 \quad \text{in} \quad \Omega \setminus N_\delta.$$

Then, if c_1/C_2 is small enough, and $\delta > 0$ is small enough, we deduce that $w \geq 0$ in $B_{1/2} \cap \Omega$.

Proof Notice that in $\Omega \setminus N_\delta$ we already know that $w > 0$. Let $y_o \in N_\delta \cap \Omega \cap B_{1/2}$, and assume by contradiction that $w(y_0) < 0$.

Consider, in $B_{1/4}(y_o)$, the function

$$v(x) = w(x) - \gamma\left\{u_{r_o}(x) - \frac{1}{2n}|x - y_o|^2\right\}.$$

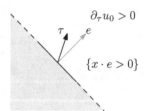

Fig. 3.8 Derivatives $\partial_\tau u_0$ are nonnegative if $\tau \cdot e \geq \frac{1}{2}$

Then, $\Delta v = 0$ in $B_{1/4}(y_0) \cap \Omega$, and $v(y_0) < 0$. Thus, v must have a negative minimum in $\partial(B_{1/4}(y_0) \cap \Omega)$.

However, if c_1/C_2 and δ are small enough, then we reach a contradiction as follows. On $\partial\Omega$ we have $v \geq 0$. On $\partial B_{1/4}(y_0) \cap N_\delta$ we have

$$v \geq -c_1 - C_0\gamma\delta^2 + \frac{\gamma}{2n}\left(\frac{1}{4}\right)^2 \geq 0 \quad \text{on} \quad \partial B_{1/4}(y_0) \cap N_\delta.$$

On $\partial B_{1/4}(y_0) \cap (\Omega \setminus N_\delta)$ we have

$$v \geq C_2 - C_0\gamma \geq 0 \quad \text{on} \quad \partial B_{1/4}(y_0) \cap (\Omega \setminus N_\delta).$$

Here, we used that $\|u_{r_0}\|_{C^{1,1}(B_1)} \leq C_0$, and chose $C_0 c_1 \leq \gamma \leq C_2/C_0$.

Using the previous lemma, we can now show that there is a cone of directions τ in which the solution is monotone near the origin.

Proposition 8 *Let u be any solution to (3.14), and assume that (3.18) holds. Let $u_r(x) = r^{-2}u(rx)$. Then, there exist $r_0 > 0$ and $e \in \mathbb{S}^{n-1}$ such that*

$$\partial_\tau u_{r_0} \geq 0 \quad \text{in} \quad B_{1/2}$$

for every $\tau \in \mathbb{S}^{n-1}$ satisfying $\tau \cdot e \geq \frac{1}{2}$.

Proof By Proposition 7, for any $\varepsilon > 0$ there exist $e \in \mathbb{S}^{n-1}$ and $r_0 > 0$ such that

$$\left|u_{r_0}(x) - \tfrac{1}{2}(x \cdot e)_+^2\right| \leq \varepsilon \quad \text{in} \quad B_1 \tag{3.20}$$

and

$$\left|\partial_\tau u_{r_0}(x) - (x \cdot e)_+(\tau \cdot e)\right| \leq \varepsilon \quad \text{in} \quad B_1 \tag{3.21}$$

for all $\tau \in \mathbb{S}^{n-1}$.

We now want to use Lemma 6 to deduce that $\partial_\tau u_{r_0} \geq 0$ if $\tau \cdot e \geq \frac{1}{2}$.

First, we claim that

$$u_{r_0} > 0 \quad \text{in} \quad \{x \cdot e > C_0\sqrt{\varepsilon}\},$$

$$u_{r_0} = 0 \quad \text{in} \quad \{x \cdot e < -C_0\sqrt{\varepsilon}\}, \tag{3.22}$$

and therefore the free boundary $\partial\Omega = \partial\{u_{r_0} > 0\}$ is contained in the strip $\{|x \cdot e| \leq C_0\sqrt{\varepsilon}\}$, for some C_0 depending only on n. To prove this, notice that if $x \cdot e > C_0\sqrt{\varepsilon}$ then

$$u_{r_0} > \frac{1}{2}(C_0\sqrt{\varepsilon})^2 - \varepsilon > 0,$$

while if there was a free boundary point $x_\circ$ in $\{x \cdot e < -C_\circ \varepsilon\}$ then by nondegeneracy we would get

$$\sup_{B_{C_\circ \sqrt{\varepsilon}}(x_\circ)} u_{r_\circ} \geq c(C_\circ \sqrt{\varepsilon})^2 > 2\varepsilon,$$

a contradiction with (3.20).

Therefore, we have

$$\partial \Omega \subset \{|x \cdot e| \leq C_\circ \sqrt{\varepsilon}\}. \tag{3.23}$$

Now, for each $\tau \in \mathbb{S}^{n-1}$ satisfying $\tau \cdot e \geq \frac{1}{2}$ we define

$$w := \partial_\tau u_{r_\circ}.$$

In order to use Lemma 6, we notice that

(a) w is bounded and harmonic in $\Omega \cap B_1$.
(b) $w = 0$ on $\partial \Omega \cap B_1$.
(c) Thanks to (3.21), if $\delta \gg \sqrt{\varepsilon}$ then w satisfies

$$w \geq -\varepsilon \quad \text{in} \quad N_\delta$$

and

$$w \geq \delta/4 > 0 \quad \text{in} \quad (\Omega \setminus N_\delta) \cap B_1.$$

(We recall $N_\delta := \{x \in B_1 : \text{dist}(x, \partial \Omega) < \delta\}$ (Fig. 3.9).)

Fig. 3.9 The setting into which we use Lemma 6

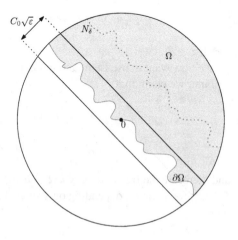

Indeed, to check the last inequality we use that, by (3.22), we have $\{x \cdot e < \delta - C_\circ\sqrt{\varepsilon}\} \subset N_\delta$. Thus, by (3.21), we get that for all $x \in (\Omega \setminus N_\delta) \cap B_1$

$$w \geq \frac{1}{2}(x \cdot e)_+ - \varepsilon \geq \frac{1}{2}\delta - \frac{1}{2}C_\circ\sqrt{\varepsilon} - \varepsilon \geq \frac{1}{4}\delta,$$

provided that $\delta \gg \sqrt{\varepsilon}$.

Using (a)–(b)–(c), we deduce from Lemma 6 that

$$w \geq 0 \quad \text{in} \quad B_{1/2}.$$

Since we can do this for every $\tau \in \mathbb{S}^{n-1}$ with $\tau \cdot e \geq \frac{1}{2}$, the proposition is proved.

As a consequence of the previous proposition, we find:

Corollary 3 *Let u be any solution to (3.14), and assume that (3.18) holds. Then, there exists $r_\circ > 0$ such that the free boundary $\partial\{u_{r_\circ} > 0\}$ is Lipschitz in $B_{1/2}$. In particular, the free boundary of u, $\partial\{u > 0\}$, is Lipschitz in $B_{r_\circ/2}$.*

Proof This follows from the fact that $\partial_\tau u_{r_\circ} \geq 0$ in $B_{1/2}$ for all $\tau \in \mathbb{S}^{n-1}$ with $\tau \cdot e \geq \frac{1}{2}$, as explained next.

Let $x_\circ \in B_{1/2} \cap \partial\{u_{r_\circ} > 0\}$ be any free boundary point in $B_{1/2}$, and let

$$\Theta := \{\tau \in \mathbb{S}^{n-1} : \tau \cdot e > \tfrac{1}{2}\},$$

$$\Sigma_1 := \{x \in B_{1/2} : x = x_\circ - t\tau, \text{ with } \tau \in \Theta,\ t > 0\},$$

and

$$\Sigma_2 := \{x \in B_{1/2} : x = x_\circ + t\tau, \text{ with } \tau \in \Theta,\ t > 0\},$$

see Fig. 3.10.

We claim that

$$\begin{aligned} u_{r_\circ} &= 0 \text{ in} \quad \Sigma_1, \\ u_{r_\circ} &> 0 \text{ in} \quad \Sigma_2. \end{aligned} \tag{3.24}$$

Indeed, since $u(x_\circ) = 0$, it follows from the monotonicity property $\partial_\tau u_{r_\circ} \geq 0$—and the nonnegativity of $u_{r_\circ}$—that $u_{r_\circ}(x_\circ - t\tau) = 0$ for all $t > 0$ and $\tau \in \Theta$. In particular, there cannot be any free boundary point in Σ_1.

On the other hand, by the same argument, if $u_{r_\circ}(x_1) = 0$ for some $x_1 \in \Sigma_2$ then we would have $u_{r_\circ} = 0$ in $\{x \in B_{1/2} : x = x_1 - t\tau, \text{ with } \tau \in \Theta,\ t > 0\} \ni x_\circ$, and in particular $x_\circ$ would not be a free boundary point. Thus, $u_{r_\circ}(x_1) > 0$ for all $x_1 \in \Sigma_2$, and (3.24) is proved.

Finally, notice that (3.24) yields that the free boundary $\partial\{u_{r_\circ} > 0\} \cap B_{1/2}$ satisfies both the interior and exterior cone condition, and thus it is Lipschitz.

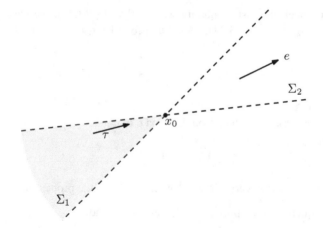

Fig. 3.10 Representation of Σ_1 and Σ_2

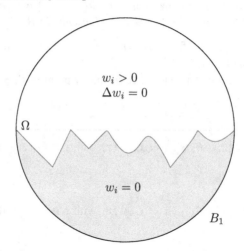

Fig. 3.11 Setting of the boundary Harnack

Once we know that the free boundary is Lipschitz, we may assume without loss of generality that $e = e_n$ and that

$$\partial\{u_{r_\circ} > 0\} \cap B_{1/2} = \{x_n = g(x')\} \cap B_{1/2}$$

for a Lipschitz function $g : \mathbb{R}^{n-1} \to \mathbb{R}$. Here, $x = (x', x_n)$, with $x' \in \mathbb{R}^{n-1}$ and $x_n \in \mathbb{R}$.

Now, we want to prove that Lipschitz free boundaries are $C^{1,\alpha}$. A key ingredient for this will be the following basic property of harmonic functions (see Fig. 3.11 for a representation of the setting).

Theorem 5 (Boundary Harnack) *Let w_1 and w_2 be positive harmonic functions in $B_1 \cap \Omega$, where $\Omega \subset \mathbb{R}^n$ is any Lipschitz domain.*

Assume that w_1 and w_2 vanish on $\partial \Omega \cap B_1$, and $C_o^{-1} \leq \|w_i\|_{L^\infty(B_{1/2})} \leq C_o$ for $i = 1, 2$. Then,

$$\frac{1}{C} w_2 \leq w_1 \leq C w_2 \qquad \text{in} \quad \overline{\Omega} \cap B_{1/2}.$$

Moreover,

$$\left\| \frac{w_1}{w_2} \right\|_{C^{0,\alpha}(\overline{\Omega} \cap B_{1/2})} \leq C$$

for some small $\alpha > 0$. The constants α and C depend only on n, C_o, and Ω.

Furthermore, if $\partial \Omega \cap B_1$ can be written as a Lipschitz graph, then C and α depend only on n, C_o, and the Lipschitz constant of Ω.

This is actually a rather difficult theorem, which does *not* follow from any explicit representation nor from Schauder-type estimates. We will not prove such theorem here; we refer to [7] for a proof of the result, as well as to [1] and [31] for a more general version of the result that allows equations with a right hand side.

Remark 3 The main point in Theorem 5 is that Ω is allowed to be *Lipschitz*. If Ω is smooth (say, C^2 or even $C^{1,\alpha}$) then it follows from a simple barrier argument that both w_1 and w_2 would be comparable to the distance to $\partial \Omega$, i.e., they vanish at a linear rate from $\partial \Omega$. However, in Lipschitz domains the result cannot be proved with a simple barrier argument, and it is much more difficult to establish.

The boundary Harnack is a crucial tool in the study of free boundary problems, and in particular in the obstacle problem. Here, we use it to prove that the free boundary is $C^{1,\alpha}$ for some small $\alpha > 0$.

Proposition 9 *Let u be any solution to (3.14), and assume that (3.18) holds. Then, there exists $r_o > 0$ such that the free boundary $\partial\{u_{r_o} > 0\}$ is $C^{1,\alpha}$ in $B_{1/4}$, for some small $\alpha > 0$. In particular, the free boundary of u, $\partial\{u > 0\}$, is $C^{1,\alpha}$ in $B_{r_o/4}$.*

Proof Let $\Omega = \{u_{r_o} > 0\}$. By Corollary 3, if $r_o > 0$ is small enough then (possibly after a rotation) we have

$$\Omega \cap B_{1/2} = \{x_n \geq g(x')\} \cap B_{1/2}$$

and the free boundary is given by

$$\partial \Omega \cap B_{1/2} = \{x_n = g(x')\} \cap B_{1/2},$$

where g is Lipschitz.

Let

$$w_2 := \partial_{e_n} u_{r_o}$$

and

$$w_1 := \partial_{e_i} u_{r_o} + \partial_{e_n} u_{r_o}, \qquad i = 1, \dots, n-1.$$

Since $\partial_\tau u_{r_o} \geq 0$ in $B_{1/2}$ for all $\tau \in \mathbb{S}^{n-1}$ with $\tau \cdot e_n \geq \frac{1}{2}$, then $w_2 \geq 0$ in $B_{1/2}$ and $w_1 \geq 0$ in $B_{1/2}$.

This is because $\partial_{e_i} + \partial_{e_n} = \partial_{e_i + e_n} = \sqrt{2}\partial_\tau$, with $\tau \cdot e_n = 1/\sqrt{2} > \frac{1}{2}$. Notice that we add the term $\partial_{e_n} u_{r_o}$ in w_1 in order to get a nonnegative function $w_2 \geq 0$.

Now since w_1 and w_2 are positive harmonic functions in $\Omega \cap B_{1/2}$, and vanish on $\partial\Omega \cap B_{1/2}$, we can use the boundary Harnack to get

$$\left\| \frac{w_1}{w_2} \right\|_{C^{0,\alpha}(\overline{\Omega} \cap B_{1/4})} \leq C$$

for some small $\alpha > 0$. Therefore, since $w_1/w_2 = 1 + \partial_{e_i} u_{r_o}/\partial_{e_n} u_{r_o}$, we deduce

$$\left\| \frac{\partial_{e_i} u_{r_o}}{\partial_{e_n} u_{r_o}} \right\|_{C^{0,\alpha}(\overline{\Omega} \cap B_{1/4})} \leq C. \tag{3.25}$$

Now, we claim that this implies that the free boundary is $C^{1,\alpha}$ in $B_{1/4}$. Indeed, if $u_{r_o}(x) = t$ then the normal vector to the level set $\{u_{r_o} = t\}$ is given by

$$\nu^i(x) = \frac{\partial_{e_i} u_{r_o}}{|\nabla u_{r_o}|} = \frac{\partial_{e_i} u_{r_o}/\partial_{e_n} u_{r_o}}{\sqrt{1 + \sum_{j=1}^n \left(\partial_{e_j} u_{r_o}/\partial_{e_n} u_{r_o} \right)^2}}, \qquad i = 1, \dots, n.$$

This is a $C^{0,\alpha}$ function by (3.25), and therefore we can take $t \to 0$ to find that the free boundary is $C^{1,\alpha}$ (since the normal vector to the free boundary is given by a $C^{0,\alpha}$ function).

So far we have proved that

$$\left(\begin{array}{c} \{u = 0\} \text{ has positive} \\ \text{density at the origin} \end{array} \right) \Longrightarrow \left(\begin{array}{c} \text{a blow-up is} \\ u_0 = \frac{1}{2}(x \cdot e)_+^2 \end{array} \right) \Longrightarrow \left(\begin{array}{c} \text{free boundary} \\ \text{is } C^{1,\alpha} \text{ near } 0 \end{array} \right)$$

As a last step in this section, we will now prove that $C^{1,\alpha}$ free boundaries are actually C^∞.

3.6.2 Higher Regularity of the Free Boundary

We want to finally prove the smoothness of free boundaries near regular points.

Theorem 6 (Smoothness of the free boundary near regular points) *Let u be any solution to (3.14), and assume that (3.18) holds. Then, the free boundary $\partial\{u > 0\}$ is C^∞ in a neighborhood of the origin.*

For this, we need the following result.

Theorem 7 (Higher order boundary Harnack) *Let $\Omega \subset \mathbb{R}^n$ be any $C^{k,\alpha}$ domain, with $k \geq 1$ and $\alpha \in (0, 1)$. Let w_1, w_2 be two solutions of $\Delta w_i = 0$ in $B_1 \cap \Omega$, $w_i = 0$ on $\partial\Omega \cap B_1$, with $w_2 > 0$ in Ω.*
Assume that $C_o^{-1} \leq \|w_i\|_{L^\infty(B_{1/2})} \leq C_o$. Then,

$$\left\| \frac{w_1}{w_2} \right\|_{C^{k,\alpha}(\overline{\Omega}\cap B_{1/2})} \leq C,$$

where C depends only on n, k, α, C_o, and Ω.

Contrary to Theorem 5, the proof of Theorem 7 is a perturbative argument, in the spirit of (but much more delicate than) Schauder estimates for linear elliptic equations. We will not prove the higher order boundary Harnack here; we refer to [12] for the proof of such result.

Using Theorem 7, we can finally give the:

Proof *(Proof of Theorem 6)* Let $u_{r_o}(x) = r_o^{-2}u(r_o x)$. By Proposition 9, we know that if $r_o > 0$ is small enough then the free boundary $\partial\{u_{r_o} > 0\}$ is $C^{1,\alpha}$ in B_1, and (possibly after a rotation) $\partial_{e_n} u_{r_o} > 0$ in $\{u_{r_o} > 0\} \cap B_1$. Thus, using the higher order boundary Harnack (Theorem 7) with $w_1 = \partial_{e_i} u_{r_o}$ and $w_2 = \partial_{e_n} u_{r_o}$, we find that

$$\left\| \frac{\partial_{e_i} u_{r_o}}{\partial_{e_n} u_{r_o}} \right\|_{C^{1,\alpha}(\overline{\Omega}\cap B_{1/2})} \leq C.$$

Actually, by a simple covering argument we find that

$$\left\| \frac{\partial_{e_i} u_{r_o}}{\partial_{e_n} u_{r_o}} \right\|_{C^{1,\alpha}(\overline{\Omega}\cap B_{1-\delta})} \leq C_\delta \tag{3.26}$$

for any $\delta > 0$.

Now, as in the proof of Proposition 9, we notice that if $u_{r_o}(x) = t$ then the normal vector to the level set $\{u_{r_o} = t\}$ is given by

$$\nu^i(x) = \frac{\partial_{e_i} u_{r_o}}{|\nabla u_{r_o}|} = \frac{\partial_{e_i} u_{r_o}/\partial_{e_n} u_{r_o}}{\sqrt{1 + \sum_{j=1}^n \left(\partial_{e_j} u_{r_o}/\partial_{e_n} u_{r_o}\right)^2}}, \qquad i = 1, \ldots, n.$$

By (3.26), this is a $C^{1,\alpha}$ function in $B_{1-\delta}$ for any $\delta > 0$, and therefore we can take $t \to 0$ to find that the normal vector to the free boundary is $C^{1,\alpha}$ inside B_1. But this means that the free boundary is actually $C^{2,\alpha}$.

Repeating now the same argument, and using that the free boundary is $C^{2,\alpha}$ in $B_{1-\delta}$ for any $\delta > 0$, we find that

$$\left\| \frac{\partial_{e_i} u_{r_\circ}}{\partial_{e_n} u_{r_\circ}} \right\|_{C^{2,\alpha}(\overline{\Omega} \cap B_{1-\delta'})} \leq C_{\delta'},$$

which yields that the normal vector is $C^{2,\alpha}$ and thus the free boundary is $C^{3,\alpha}$. Iterating this argument, we find that the free boundary $\partial\{u_{r_\circ} > 0\}$ is C^∞ inside B_1, and hence $\partial\{u > 0\}$ is C^∞ in a neighborhood of the origin.

This completes the study of *regular* free boundary points. It remains to understand what happens at points where the contact set has *density zero* (see e.g. Fig. 3.5). This is the content of the next section.

3.7 Singular Points

We finally study the behavior of the free boundary at singular points, i.e., when

$$\lim_{r \to 0} \frac{\left|\{u = 0\} \cap B_r\right|}{|B_r|} = 0. \tag{3.27}$$

For this, we first notice that, as a consequence of the results of the previous section, we get the following.

Proposition 10 *Let u be any solution to (3.14). Then, we have the following dichotomy:*

(a) *Either (3.18) holds and all blow-ups of u at 0 are of the form*

$$u_0(x) = \frac{1}{2}(x \cdot e)_+^2,$$

for some $e \in \mathbb{S}^{n-1}$.

(b) *Or (3.27) holds and all blow-ups of u at 0 are of the form*

$$u_0(x) = \frac{1}{2}x^T A x,$$

for some matrix $A \geq 0$ with $\operatorname{tr} A = 1$.

Points of the type (a) were studied in the previous section; they are called *regular* points and the free boundary is C^∞ around them (in particular, the blow-up is unique). Points of the type (b) are those at which the contact set has zero density, and are called *singular* points.

To prove the result, we need the following:

Lemma 7 *Let u be any solution to (3.14), and assume that (3.27) holds. Then, every blow-up of u at 0 satisfies $|\{u_0 = 0\}| = 0$.*

Proof Let u_0 be a blow-up of u at 0, i.e., $u_{r_k} \to u_0$ in $C^1_{\mathrm{loc}}(\mathbb{R}^n)$ along a sequence $r_k \to 0$, where $u_r(x) = r^{-2}u(rx)$.

Notice that the functions u_r solve

$$\Delta u_r = \chi_{\{u_r > 0\}} \quad \text{in} \quad B_1,$$

in the sense that

$$\int_{B_1} \nabla u_r \cdot \nabla \eta \, dx = \int_{B_1} \chi_{\{u_r > 0\}} \eta \, dx \qquad \text{for all } \eta \in C_c^\infty(B_1). \tag{3.28}$$

Moreover, by assumption (3.27), we have $\left|\{u_r = 0\} \cap B_1\right| \longrightarrow 0$, and thus taking limits $r_k \to 0$ in (3.28) we deduce that $\Delta u_0 = 1$ in B_1. Since we know that u_0 is convex, nonnegative, and homogeneous, this implies that $|\{u_0 = 0\}| = 0$. $\qquad\blacksquare$

We can now give the:

Proof (*Proof of Theorem* 10) By the classification of blow-ups, Theorem 4, the possible blow-ups can only have one of the two forms presented. If (3.18) holds for at least one blow-up, thanks to the smoothness of the free boundary (by Proposition 9), it holds for all blow-ups, and thus, by Corollary 2, $u_0(x) = \frac{1}{2}(x \cdot e)_+^2$ (and in fact, the smoothness of the free boundary yields uniqueness of the blow-up in this case).

If (3.27) holds, then by Lemma 7 the blow-up u_0 must satisfy $\left|\{u_0 = 0\}\right| = 0$, and thus we are in case (b) (see the proof of Theorem 4). $\qquad\blacksquare$

In the previous section we proved that the free boundary is C^∞ in a neighborhood of any regular point. A natural question then is to understand better the solution u near singular points. One of the main results in this direction is the following.

Theorem 8 (Uniqueness of blow-ups at singular points) *Let u be any solution to (3.14), and assume that 0 is a singular free boundary point.*

Then, there exists a homogeneous quadratic polynomial $p_2(x) = \frac{1}{2}x^T A x$, with $A \geq 0$ and $\Delta p_2 = 1$, such that

$$u_r \longrightarrow p_2 \quad \text{in} \quad C^1_{\mathrm{loc}}(\mathbb{R}^n).$$

In particular, the blow-up of u at 0 is unique, and $u(x) = p_2(x) + o(|x|^2)$.

To prove this, we need the following monotonicity formula due to Monneau.

Theorem 9 (Monneau's monotonicity formula) *Let u be any solution to* (3.14), *and assume that* 0 *is a singular free boundary point.*

Let q be any homogeneous quadratic polynomial with $q \geq 0$, $q(0) = 0$, *and* $\Delta q = 1$. *Then, the quantity*

$$M_{u,q}(r) := \frac{1}{r^{n+3}} \int_{\partial B_r} (u - q)^2$$

is monotone in r, that is, $\frac{d}{dr} M_{u,q}(r) \geq 0$.

Proof We sketch the argument here, and refer to [28, Theorem 7.4] for more details. We first notice that

$$M_{u,q}(r) = \int_{\partial B_1} \frac{(u - q)^2(rx)}{r^4},$$

and hence a direct computation yields

$$\frac{d}{dr} M_{u,q}(r) = \frac{2}{r^{n+4}} \int_{\partial B_r} (u - q) \{x \cdot \nabla(u - q) - 2(u - q)\}.$$

On the other hand, it turns out that

$$\frac{1}{r^{n+3}} \int_{\partial B_r} (u - q) \{x \cdot \nabla(u - q) - 2(u - q)\} = W_u(r) - W_u(0^+) +$$

$$+ \frac{1}{r^{n+2}} \int_{B_r} (u - q)\Delta(u - q),$$

where $W_u(r)$ (as defined in (3.15)) is monotone increasing in $r > 0$ thanks to Theorem 3. Thus, we have

$$\frac{d}{dr} M_{u,q}(r) \geq \frac{2}{r^{n+3}} \int_{B_r} (u - q)\Delta(u - q).$$

But since $\Delta u = \Delta q = 1$ in $\{u > 0\}$, and $(u - q)\Delta(u - q) = q \geq 0$ in $\{u = 0\}$, then we have

$$\frac{d}{dr} M_{u,q}(r) \geq \frac{2}{r^{n+3}} \int_{B_r \cap \{u=0\}} q \geq 0,$$

as wanted.

We can now give the:

Proof *(Proof of Theorem 8)* By Proposition 10 (and Proposition 6), we know that at any singular point we have a subsequence $r_j \to 0$ along which $u_{r_j} \to p$ in

$C^1_{loc}(\mathbb{R}^n)$, where p is a 2-homogeneous quadratic polynomial satisfying $p(0) = 0$, $p \geq 0$, and $\Delta p = 1$. Thus, we can use Monneau's monotonicity formula with such polynomial p to find that

$$M_{u,p}(r) := \frac{1}{r^{n+3}} \int_{\partial B_r} (u - p)^2$$

is monotone increasing in $r > 0$. In particular, the limit $\lim_{r \to 0} M_{u,p}(r) =: M_{u,p}(0^+)$ exists.

Now, recall that we have a sequence $r_j \to 0$ along which $u_{r_j} \to p$. In particular, $r_j^{-2} \{u(r_j x) - p(r_j x)\} \longrightarrow 0$ locally uniformly in $\mathbb{R}^n$, i.e.,

$$\frac{1}{r_j^2} \|u - p\|_{L^\infty(B_{r_j})} \longrightarrow 0$$

as $r_j \to 0$. This yields that

$$M_{u,p}(r_j) \leq \frac{1}{r_j^{n+3}} \int_{\partial B_{r_j}} \|u - p\|^2_{L^\infty(B_{r_j})} \longrightarrow 0$$

along the subsequence $r_j \to 0$, and therefore $M_{u,p}(0^+) = 0$.

Let us show that this implies the uniqueness of blow-up. Indeed, if there was another subsequence $r_\ell \to 0$ along which $u_{r_\ell} \to q$ in $C^1_{loc}(\mathbb{R}^n)$, for a 2-homogeneous quadratic polynomial q, then we would repeat the argument above to find that $M_{u,q}(0^+) = 0$. But then this yields, by homogeneity of p and q,

$$\int_{\partial B_1} (p - q)^2 = \frac{1}{r^{n+3}} \int_{\partial B_r} (p - q)^2 \leq 2M_{u,p}(r) + 2M_{u,q}(r) \longrightarrow 0,$$

and hence

$$\int_{\partial B_1} (p - q)^2 = 0.$$

This means that $p = q$, and thus the blow-up of u at 0 is unique.

Let us finally show that $u(x) = p(x) + o(|x|^2)$, i.e., $r^{-2}\|u - p\|_{L^\infty(B_r)} \to 0$ as $r \to 0$. Indeed, assume by contradiction that there is a subsequence $r_k \to 0$ along which

$$r_k^{-2}\|u - p\|_{L^\infty(B_{r_k})} \geq c_1 > 0.$$

Then, there would be a subsequence of r_{k_i} along which $u_{r_{k_i}} \to u_0$ in $C^1_{loc}(\mathbb{R}^n)$, for a certain blow-up u_0 satisfying $\|u_0 - p\|_{L^\infty(B_1)} \geq c_1 > 0$. However, by uniqueness of blow-up it must be $u_0 = p$, and hence we reach a contradiction.

Summarizing, we have proved the following result:

Theorem 10 *Let u be any solution to* (3.14). *Then, we have the following dichotomy:*

(a) *Either all blow-ups of u at 0 are of the form*

$$u_0(x) = \frac{1}{2}(x \cdot e)_+^2 \qquad \text{for some} \quad e \in \mathbb{S}^{n-1},$$

and the free boundary is C^∞ *in a neighborhood of the origin.*

(b) *Or there is a homogeneous quadratic polynomial p, with* $p(0) = 0$, $p \geq 0$, *and* $\Delta p = 1$, *such that*

$$\|u - p\|_{L^\infty(B_r)} = o(r^2) \qquad \text{as} \quad r \to 0.$$

In particular, when this happens we have

$$\lim_{r \to 0} \frac{|\{u = 0\} \cap B_r|}{|B_r|} = 0.$$

The last question that remains to be answered is: How large can the set of singular points be? This is the topic of the following section.

3.8   The Size of the Singular Set

We finish these notes with a discussion of more recent results (as well as some open problems) about the set of singular points.

Recall that a free boundary point $x_\circ \in \partial\{u > 0\}$ is singular whenever

$$\lim_{r \to 0} \frac{|\{u = 0\} \cap B_r(x_\circ)|}{|B_r(x_\circ)|} = 0.$$

The main known result on the size of the singular set reads as follows.

Theorem 11 ([6]) *Let u be any solution to* (3.14). *Let* $\Sigma \subset B_1$ *be the set of singular points.*

Then, $\Sigma \cap B_{1/2}$ *is contained in a* C^1 *manifold of dimension* $n - 1$.

This result is sharp, in the sense that it is not difficult to construct examples in which the singular set is $(n - 1)$-dimensional; see [32].

As explained below, such result essentially follows from the uniqueness of blow-ups at singular points, established in the previous section.

Indeed, given any singular point $x_\circ$, let $p_{x_\circ}$ be the blow-up of u at $x_\circ$ (recall that $p_{x_\circ}$ is a nonnegative 2-homogeneous polynomial). Let k be the dimension of

Fig. 3.12 u is positive in
$\{x \in B_\rho(x_o) :$
$|(x - x_o) \cdot e_{x_o}| > \omega(|x - x_o|)\}$

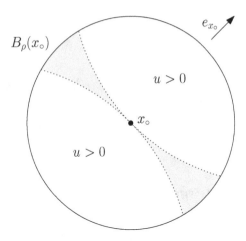

the set $\{p_{x_o} = 0\}$—notice that this is a proper linear subspace of $\mathbb{R}^n$, so that $k \in \{0, \ldots, n-1\}$—and define

$$\Sigma_k := \big\{x_o \in \Sigma : \dim(\{p_{x_o} = 0\}) = k\big\}. \tag{3.29}$$

Clearly, $\Sigma = \cup_{k=0}^{n-1} \Sigma_k$.

The following result gives a more precise description of the singular set.

Proposition 11 ([6]) *Let u be any solution to (3.14). Let $\Sigma_k \subset B_1$ be defined by (3.29), $k = 1, \ldots, n-1$. Then, $\Sigma_k \cap B_{1/2}$ is contained in a C^1 manifold of dimension k.*

The rough heuristic idea of the proof of this result is as follows. Assume for simplicity that $n = 2$, so that $\Sigma = \Sigma_1 \cup \Sigma_0$.

Let us take a point $x_o \in \Sigma_0$. Then, by Theorem 10, we have the expansion

$$u(x) = p_{x_o}(x - x_o) + o\big(|x - x_o|^2\big) \tag{3.30}$$

where p_{x_o} is the blow-up of u at x_o (recall that this came from the uniqueness of blow-up at x_o). By definition of Σ_0, the polynomial p_{x_o} must be positive outside the origin, and thus by homogeneity satisfies $p_{x_o}(x - x_o) \geq c|x - x_o|^2$, with $c > 0$. This, combined with (3.30), yields then that u must be positive in a neighborhood of the origin. In particular, all points in Σ_0 are isolated.

On the other hand, let us now take a point $x_o \in \Sigma_1$. Then, by definition of Σ_1 the blow-up must necessarily be of the form $p_{x_o}(x) = \frac{1}{2}(x \cdot e_{x_o})^2$, for some $e_{x_o} \in \mathbb{S}^{n-1}$. Again by the expansion (3.30), we find that u is positive in a region of the form

$$\big\{x \in B_\rho(x_o) : |(x - x_o) \cdot e_{x_o}| > \omega(|x - x_o|)\big\},$$

where ω is a certain modulus of continuity, and $\rho > 0$ is small (see Fig. 3.12).

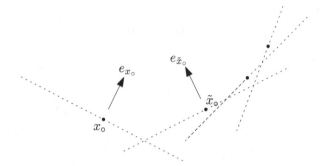

Fig. 3.13 Singular points $x_\circ, \tilde{x}_\circ \in \Sigma_1$

This is roughly saying that the set Σ_1 "has a tangent plane" at $x_\circ$. Repeating the same at any other point $\tilde{x}_\circ \in \Sigma_1$ we find that the same happens at every point in Σ_1 and, moreover, if $\tilde{x}_\circ$ is close to $x_\circ$ then $e_{\tilde{x}_\circ}$ must be close to $e_{x_\circ}$—otherwise the expansions (3.30) at $\tilde{x}_\circ$ and $x_\circ$ would not match. Finally, since the modulus ω can be made independent of the point (by a compactness argument), then it turns out that the set Σ_1 is contained in a C^1 curve (see Fig. 3.13).

What we discussed here is just an heuristic argument; the actual proof uses Whitney's extension theorem and can be found for example in [28].

We finally discuss some recent results [17] on the fine structure and regularity of singular points.

3.9   Finer Understanding of Singular Points

In order to get a finer understanding of singular points, we will need to relate the obstacle problem to the so-called *thin obstacle problem*. For this, we first briefly introduce such free boundary problem and summarize the main known results in this context.

3.9.1   The Thin Obstacle Problem

The thin obstacle problem is another classical free boundary problem, which was originally studied by Signorini in connection with linear elasticity [34, 35]. The problem gained further attention in the seventies due to its connection to mechanics, biology, and even finance—see [9, 14, 25], and [30]—, and since then it has been widely studied in the mathematical community; see [2–5, 8, 10, 15, 19, 21, 23, 28, 33] and references therein.

We say that $w \in H^1(B_1)$ is a solution to the thin obstacle problem (with zero obstacle) if

$$\begin{cases} -\Delta w = 0 & \text{in } B_1 \setminus (\{x_n = 0\} \cap \{w = 0\}) \\ -\Delta w \geq 0 & \text{in } B_1 \\ w \geq 0 & \text{on } \{x_n = 0\} \\ w = g & \text{on } \partial B_1, \end{cases} \tag{3.31}$$

in the weak sense, for some boundary data $g \in C^0(\partial B_1 \cap \{x_n \geq 0\})$. These solutions are minimizers of the Dirichlet energy

$$\int_{B_1} |\nabla w|^2,$$

under the constrain $w \geq 0$ on $\{x_n = 0\}$, and with boundary conditions $w = g$ on ∂B_1.

The contact set is denoted by

$$\Lambda(w) := \{x' \in \mathbb{R}^{n-1} : w(x', 0) = 0\},$$

and the free boundary is $\Gamma(w) = \partial \Lambda(w)$. Here, we denoted $x = (x', x_n) \in \mathbb{R}^{n-1} \times \mathbb{R}$.

Solutions to (3.31) are $C^{1,\frac{1}{2}}$ (see [2]), and this is optimal.

A key tool in establishing the optimal regularity of solutions is the Almgren frequency formula:

$$N_w(r) := \frac{r \int_{B_r} |\nabla w|^2}{\int_{\partial B_r} w^2} \qquad \text{is monotone in } r, \tag{3.32}$$

namely, $N'_w(r) \geq 0$ for $r \in (0, 1)$.

This allows one to show that *blow-ups are homogeneous*, and the degree of homogeneity of any blow-up of w at 0 is exactly $N_w(0^+)$. Moreover, it was shown in [3] that the lowest possible homogeneity is in fact $3/2$, and this gives the optimal $C^{1,\frac{1}{2}}$ regularity of solutions; we refer to [3] for more details.

The Free Boundary in the Thin Obstacle Problem

The main known results concerning the structure of the free boundary are the following.

The free boundary can be divided into two sets,

$$\Gamma(w) = \text{Reg}(w) \cup \text{Deg}(w),$$

the set of *regular points*,

$$\text{Reg}(w) := \left\{ x \in \Gamma(w) : 0 < cr^{3/2} \le \sup_{B_r(x)} w \le Cr^{3/2}, \quad \forall r \in (0, r_o) \right\},$$

and the set of non-regular points or *degenerate points*

$$\text{Deg}(w) := \left\{ x \in \Gamma(w) : 0 \le \sup_{B_r(x)} w \le Cr^2, \quad \forall r \in (0, r_o) \right\}, \tag{3.33}$$

Alternatively, each of the subsets can be defined according to the homogeneity of the blow-up at that point. Indeed, the set of regular points are those whose blow-up is of order $\frac{3}{2}$, and the set of degenerate points are those whose blow-up is of order κ for some $\kappa \in [2, \infty)$.

Let us denote Γ_κ the set of free boundary points of order κ. That is, those points whose blow-up is homogeneous of order κ. Then, the free boundary can be divided as

$$\Gamma(w) = \Gamma_{3/2} \cup \Gamma_{\text{even}} \cup \Gamma_{\text{odd}} \cup \Gamma_{\text{half}} \cup \Gamma_*, \tag{3.34}$$

where:

- $\Gamma_{3/2} = \text{Reg}(w)$ is the set of regular points. They are an open $(n-2)$-dimensional subset of $\Gamma(w)$, and it is C^∞ (see [3] and [13, 23]).
- $\Gamma_{\text{even}} = \bigcup_{m \ge 1} \Gamma_{2m}(w)$ denotes the set of points whose blow-ups have even homogeneity. Equivalently, they can also be characterised as those points of the free boundary where the contact set has zero density, and they are often called singular points. They are contained in the countable union of C^1 $(n-2)$-dimensional manifolds; see [21].
- $\Gamma_{\text{odd}} = \bigcup_{m \ge 1} \Gamma_{2m+1}(w)$ is also an at most $(n-2)$-dimensional subset of the free boundary and it is $(n-2)$-rectifiable (see [19]). They can be characterised as those points of the free boundary where the contact set has density one.
- $\Gamma_{\text{half}} = \bigcup_{m \ge 1} \Gamma_{2m+3/2}(w)$ corresponds to those points with blow-up of order $\frac{7}{2}$, $\frac{11}{2}$, etc. They are much less understood than regular points. The set Γ_{half} is an $(n-2)$-dimensional subset of the free boundary and it is a $(n-2)$-rectifiable set (see [19]).
- Γ_* is the set of all points with homogeneities $\kappa \in (2, \infty)$, with $\kappa \notin \mathbb{N}$ and $\kappa \notin 2\mathbb{N} - \frac{1}{2}$. This set has Hausdorff dimension at most $n - 3$, so it is always *small*.

Remark 4 It is interesting to notice that, if w solves the Signorini problem (3.31), then $U(x) = \{w, -w\}$ is a $C^{1,\mu}$ two-valued harmonic function, to which the results of [24] apply. In particular, most of the above results for (3.31) follow from [24].

Dimension-Reduction in the Thin Obstacle Problem

Dimension-reduction arguments were first introduced by Almgren in the context of minimal surfaces, and are nowadays used in a variety of settings in PDEs. We will later describe some recent results [17] on the obstacle problem that are based on such kind of arguments.

Now, in order to present these ideas in a simpler context, we will show the following result for the thin obstacle problem:

Theorem 12 *Let w be any solution to the thin obstacle problem. Then,*

$$\dim_{\mathcal{H}}(\Gamma_*) \leq n - 3.$$

In order to prove Theorem 12, we follow the arguments of [37]. A key tool is the Almgren frequency function

$$N_{w,x}(r) := \frac{r \int_{B_r(x)} |\nabla w|^2}{\int_{\partial B_r(x)} w^2}, \tag{3.35}$$

which is monotone in r. In particular, we can define

$$N_{w,x}(0+) := \lim_{r \to 0} N_{w,x}(r).$$

It is not difficult to show that such quantity is upper semicontinuous in x, that is, if $w_j \to w$ uniformly in B_1 and $x_j \to x$ then

$$\limsup_j N_{w_j, x_j}(0+) \leq N_{w,x}(0+).$$

Notice also that, if q is any global μ-homogeneous solution of the thin obstacle problem, then:

- $N_{q,0}(r) \equiv \mu$ is constant in r.
- $N_{q,x}(0+) \leq \mu$ for all $x \in \{x_n = 0\}$.
- The set $S(q) = \{x : N_{q,x}(0+) = \mu\}$ is a linear subspace, and $q(x + y) = q(y)$ for all $y \in \mathbb{R}^n$ and $x \in S(q)$.
- If $\mu \notin \mathbb{N}$ and $\mu \notin 2\mathbb{N} - \frac{1}{2}$, then $S(q)$ is of dimension at most $n - 3$.

The last property follows simply from the fact that the only possible homogeneities μ for the thin obstacle problem in dimension $n = 2$ are $\mu = \frac{3}{2}, \frac{7}{2}, \frac{11}{2}, \ldots$ or $\mu = 2, 3, 4, 5, \ldots$; see for example [21].

Using this, we will now prove:

Lemma 8 *Let w be any solution to the thin obstacle problem and $x \in \Gamma_*$.*
Then, for every $\delta > 0$ there exists $\varepsilon > 0$ and $\rho > 0$ (depending on u, x, δ) such that for every $r \in (0, \rho]$ there exists an $(n - 3)$-dimensional plane $L_{x,r} \subset \{x_n = 0\}$,

passing through x, such that

$$B_r(x) \cap \{y : N_{w,y}(0+) \geq N_{w,x}(0+) - \varepsilon\} \subset \{y : \text{dist}(y, L_{x,r}) < \delta r\}.$$

Proof We prove the result for $x = 0$. Denote

$$w_r(x) := \frac{w(rx)}{\left(\fint_{\partial B_r} w^2\right)^{1/2}}.$$

Recall that w_r converges along subsequences to homogeneous blow-ups q, which are homogeneous of degree $\mu = N_{w,x}(0+)$. (Notice that a priori different subsequences could lead to different blow-ups, but they all have the same homogeneity.)

Assume by contradiction that for some $\delta > 0$, $\varepsilon_k \downarrow 0$, and $r_k \downarrow 0$, we have

$$B_{r_k} \cap \{y : N_{w,y}(0+) \geq \mu - \varepsilon_k\} \not\subset \{y : \text{dist}(y, L) < \delta r_k\}$$

for every $(n - 3)$-dimensional linear subspace L of $\{x_n = 0\}$. By scaling,

$$B_1 \cap \{y : N_{w_{r_k}, y}(0+) \geq \mu - \varepsilon_k\} \not\subset \{y : \text{dist}(y, L) < \delta\} \tag{3.36}$$

for every $(n - 3)$-dimensional linear subspace L of $\{x_n = 0\}$. By $C^{1,\alpha}$ estimates and the Arzela-Ascoli theorem, after passing to a subsequence, w_{r_k} converges to a blow-up q in the C^1 norm on each compact subset of $\{x_n \geq 0\}$. Since $0 \in \Gamma_*$, then $\dim S(q) \leq n - 3$, so we can take $L = S(q)$ (or any $(n - 3)$-dimensional linear subspace containing $S(q)$).

By (3.36), there exists z_k with $N_{w_{r_k}, z_k}(0+) \geq \mu - \varepsilon_k$ and $\text{dist}(z_k, L) \geq \delta$. After passing to a subsequence, $z_k \to z$. By the upper semicontinuity of frequency, $N_{q,z}(0+) \geq \mu$, implying that $z \in S(q)$. But $\text{dist}(z, S(q)) \geq \text{dist}(z, L) \geq \delta$, a contradiction.

We will also need the following result, whose proof can be found for example in [37].

Proposition 12 *Let $E \subseteq \mathbb{R}^n$ such that for each $\delta > 0$, each $x \in E$, and each $r > 0$ small enough, there exists an m-dimensional plane $L_{x,r}$, passing through x, for which*

$$E \cap B_r(x) \subset \{y : \text{dist}(y, L_{x,r}) < \delta r\}.$$

Then, $\dim_{\mathcal{H}}(E) \leq m$.

For convenience of the reader, we provide a proof of Proposition 12 in the Appendix.

Using this, we can give the:

Proof *(Proof of Theorem 12)* Let $\delta > 0$. For $i = 1, 2, 3, \ldots$, define $\Gamma_*^{(i)}$ to be the set of all points $y \in \Gamma_*$ such that the conclusion of Lemma 8 holds true with $\varepsilon = 1/i$ and $\rho = 1/i$. Observe that $\Gamma_* = \bigcup_i \Gamma_*^{(i)}$. For each $j = 1, 2, 3, \ldots$, define

$$\Gamma_*^{(i,j)} = \{x \in \Gamma_*^{(i)} : (j-1)/i < N_{w,x}(0+) \leq j/i\}.$$

Observe that $\Gamma_* = \bigcup_{i,j} \Gamma_*^{(i,j)}$. Moreover, for every $x \in \Gamma_*^{(i,j)}$,

$$\Gamma_*^{(i,j)} \subset \{y : N_{w,y}(0+) > N_{w,x}(0+) - 1/i\}$$

and thus by Lemma 8 for every $r \in (0, 1/i]$ there exists a $(n-3)$-dimensional plane $L_{x,r}$ of $\{x_n = 0\}$, passing through x, such that

$$\Gamma_*^{(i,j)} \cap B_r(x) \subset \{y : \text{dist}(y, L_{x,r}) < \delta r\}.$$

Since $\delta > 0$ is arbitrary, by Proposition 12 with $E = \Gamma_*^{(i,j)}$, we have $\dim_{\mathcal{H}}(\Gamma_*^{(i,j)}) \leq m$, and thus Γ_* has Hausdorff dimension at most m.

3.9.2   Relating the Obstacle Problem and the Thin Obstacle Problem

A key idea in [17] is to notice that, if u is a solution to the obstacle problem, 0 is a singular point, and we consider

$$w = u - p_2,$$

where p_2 is the blow-up of u at 0, then w behaves like a solution to the thin obstacle problem.

Indeed, since $\Delta p_2 = 1$ then $\Delta w = -\chi_{\{u=0\}}$, and therefore w solves

$$\begin{cases} -\Delta w = 0 & \text{in } B_1 \setminus \{u = 0\} \\ -\Delta w \geq 0 & \text{in } B_1 \\ \quad\; w \geq 0 & \text{on } \{p_2 = 0\}. \end{cases}$$

When p_2 is of the form $p_2(x) = \frac{1}{2}(x_n)^2$, then as we rescale w closer and closer to the origin, it turns out that $\{u = 0\}$ becomes closer and closer to $\{p_2 = 0\} = \{x_n = 0\}$, and thus w becomes closer and closer to a solution to the Signorini problem (or simply an harmonic function if $\{p_2 = 0\}$ is too small).

To make this argument precise, we need the following.

Proposition 13 *Let u be a solution to the obstacle problem in B_1, and assume that 0 is a singular point. Let p_2 be the blow-up of u at 0, and let*

$$w := u - p_2$$

and

$$N_w(r) := \frac{r \int_{B_r} |\nabla w|^2}{\int_{\partial B_r} w^2}. \tag{3.37}$$

Then for all $r \in (0, 1)$ we have $N_w(r) \geq 2$ and

$$N_w'(r) \geq \frac{2}{r} \frac{\left(r \int_{B_r} w \Delta w\right)^2}{\left(\int_{\partial B_r} w^2\right)^2} \geq 0, \quad \forall r \in (0, 1).$$

Proof Let

$$H(r) = r^{1-n} \int_{\partial B_r} w^2, \quad D(r) = r^{2-n} \int_{\partial B_r} |\nabla w|^2.$$

Then, we have

$$H'(1) = 2 \int_{\partial B_1} w w_\nu$$

$$D'(1) = 2 \int_{\partial B_1} w_\nu^2 - \int_{B_1} (x \cdot \nabla w) \Delta w$$

$$D(r) = \int_{\partial B_1} w w_\nu - \int_{B_1} w \Delta w$$

We first claim that

$$D(r) \geq 2H(r).$$

Indeed, thanks to Weiss' monotonicity formula (Theorem 3) we have

$$0 \le W_u(r) - W_u(0^+) = W_u(r) - W_{p_2}(r)$$

$$= (\text{Exercise})$$

$$= r^{-2-n} \int_{\partial B_r} |\nabla w|^2 - 2r^{-3-n} \int_{\partial B_r} w^2$$

$$= \frac{1}{r^4}(D(r) - 2H(r)),$$

and thus the claim follows.

On the other hand, since

$$N_w(r) = \frac{D(r)}{H(r)}$$

we then have

$$N'_w(1) = \frac{D'(1)H(1) - H'(1)D(1)}{H(1)^2}$$

$$= \frac{2\left(\int_{\partial B_1} w_\nu^2 - \int_{B_1}(x \cdot \nabla w)\Delta w\right)\int_{\partial B_1} w^2 - 2\int_{\partial B_1} ww_\nu\left(\int_{\partial B_1} ww_\nu - \int_{B_1} w\Delta w\right)}{H(1)^2}$$

$$= 2\frac{\int_{\partial B_1} w_\nu^2 \int_{\partial B_1} w^2 - \left(\int_{\partial B_1} ww_\nu\right)^2 + \text{rest}}{H(1)}.$$

where

$$\text{rest} := \left(\int_{B_1} w\Delta w\right)\left(\int_{\partial B_1} ww_\nu\right) - \left(\int_{B_1}(x \cdot \nabla w)\Delta w\right)\int_{\partial B_1} w^2$$

$$= \left(\int_{B_1} w\Delta w\right)^2 + \left(\int_{B_1} w\Delta w\right)D(1) - \left(\int_{B_1}(x \cdot \nabla w)\Delta w\right)H(1).$$

But now recall that $\Delta w = -\chi_{\{u=0\}}$ and $(x \cdot \nabla w) = -2x \cdot \nabla p = -2p$ on $\{u = 0\}$ thus we obtain

$$\int_{B_1} w\Delta w = \int_{\{u=0\}\cap B_1} p \ge 0 \qquad \int_{B_1}(x \cdot \nabla w)\Delta w = 2\int_{\{u=0\}\cap B_1} p \ge 0$$

Using this we get

$$\text{rest} = \left(\int_{B_1} w\Delta w\right)^2 + \left(D(1) - 2H(1)\right) \int_{\{u=0\}\cap B_1} p \ge 0,$$

where we used that $D(1) \ge 2H(1)$.

The previous proposition allows us to show the following. Recall that the sets Σ_k were defined in (3.29).

Corollary 4 *Let u be a solution to the obstacle problem in B_1, and assume that 0 is a singular point. Let p_2 be the blow-up of u at 0, and let*

$$w := u - p_2$$

and

$$w_r(x) := \frac{w(rx)}{r^{(1-n)/2}\|w\|_{L^2(\partial B_r)}}. \tag{3.38}$$

Then, $w_{r_j} \to q$ in $L^2_{\text{loc}}(\mathbb{R}^n)$ along a subsequence $r_j \to 0$, and:

- *If $0 \in \Sigma_{n-1}$, then we have that q is a homogeneous solution to the Signorini problem of degree >2.*
- *If $0 \in \Sigma_k$, $k \le n - 2$, then we have that q is a homogeneous harmonic function of degree ≥ 2.*

Moreover, we also have the following.

Proposition 14 *Let u be a solution to the obstacle problem in B_1, and assume that 0 is a singular point. Let p_2 be the blow-up of u at 0, and let*

$$w := u - p_2$$

and $N_w(r)$ be given by (3.37).
   Assume that $N_w(0^+) \ge \kappa$. Then, the quantity

$$\frac{1}{r^{n-1+2\kappa}} \int_{\partial B_r} (u - p_2)^2$$

is monotone nondecreasing for $r \in (0, 1)$.

Proof Denote, as before,

$$H(r) = r^{1-n} \int_{\partial B_r} w^2.$$

Then,

$$\frac{d}{dr}\left\{\frac{1}{r^{2\kappa}}H(r)\right\} = \frac{1}{r^{2\kappa+1}}\left(r H'(r) - 2\kappa\, H(r)\right).$$

Now, using that

$$N_w(r) = \frac{D(r)}{H(r)} = \frac{\frac{1}{2}r H'(r) - r^{2-n}\int_{B_r} w\Delta w}{H(r)},$$

that $w\Delta w \geq 0$, and that $N_w(r)$ is monotone, we have

$$\kappa \leq N_w(0^+) \leq N_w(r) \leq \frac{\frac{1}{2}r H'(r)}{H(r)}.$$

This yields $r H'(r) \geq 2\kappa\, H(r)$, and the result follows.

Notice that the previous result is an improved Monneau-type monotonicity formula, which gives a finer information at singular points whenever $\kappa > 2$.

Finally, it was proved in [17] with a dimension reduction argument that the following holds.

Theorem 13 *Let u be a solution to the obstacle problem in B_1. Let p_{2,x_0} be the blow-up of u at x_0, and let*

$$w_{x_0}(x) := u(x_0 + x) - p_{2,x_0}(x)$$

and

$$N_{w,x_0}(r) := \frac{r\int_{B_r}|\nabla w_{x_0}|^2}{\int_{\partial B_r} w_{x_0}^2}.$$

Then, outside a set of Hausdorff dimension $n - 3$, we have $N_{w,x_0}(0^+) \geq 3$.

Notice that the analysis is quite different in Σ_{n-1} or in Σ_k for $k \leq n - 2$. In the first case, one must show that the set of points $x_0 \in \Sigma_{n-1}$ at which the blow-up of w_{x_0} is a solution to the thin obstacle problem with homogeneity in the interval $(2, 3)$ is small. This is analogous to what happens in Theorem 12. In the second case, instead, one must show that the set of points $x_0 \in \Sigma_k$ at which the blow-up of w_{x_0} is homogeneous of degree 2 is small, and the corresponding argument is different.

As a consequence of Theorem 13, it was shown in [17] that the singular set is actually contained in a $C^{1,1}$ manifold, outside a (relatively open) set of Hausdorff dimension $n - 3$.

Finally, an alternative way to state the above result is that, outside a set of Hausdorff dimension $n - 3$, we have

$$u(x) = p_{2,x_0}(x - x_0) + O(|x - x_0|^3).$$

This gives the sharp rate of convergence of blow-ups for the obstacle problem in $\mathbb{R}^n$; we refer to [17, 18] for more details.

Appendix: Proof of Proposition 12

Proposition 12 will follow from the following.

Lemma 9 *For every $\beta > 0$ there exists $\delta > 0$ such that the following holds.*
Let $E \subseteq \mathbb{R}^n$ such that for each $x \in E$ and $r \in (0, r_0)$ there exists a m-dimensional plane $L_{x,r}$, passing through x, for which

$$E \cap B_r(x) \subset \{y : \text{dist}(y, L_{x,r}) < \delta r\}.$$

Then, $\mathcal{H}^{m+\beta}(A) = 0$.

Proof By a covering argument, we may assume that $E \subseteq B_1$ and $0 \in E$. By assumption, there exists a plane $L_{0,1}$ such that

$$E \cap B_1 \subset \{y : \text{dist}(y, L_{0,1}) < \delta\}.$$

Cover $L_{0,1}$ by a finite collection of balls $\{B_{2\delta}(z_k)\}_{k=1,2,\dots,N}$ where $z_k \in L_{0,1}$ for each k and $N \leq C\delta^{-m}$. Observe that $\{B_{2\delta}(z_k)\}_{k=1,2,\dots,N}$ covers $\{y : \text{dist}(y, L_{0,1}) < \delta\}$ and thus covers $E \cap B_1$. Throw away the balls $B_{2\delta}(z_k)$ that do not intersect E. For the remaining balls, let $x_k \in A \cap B_{2\delta}(z_k)$. Now $\{B_{4\delta}(x_k)\}_{k=1,2,\dots,N}$ covers $E \cap B_1$, $x_k \in E$, with $N \leq C\delta^{-m}$, and thus $N(4\delta)^{m+\beta} \leq C\delta^\beta \leq 1/2$, provided that $\delta > 0$ is small enough.

Now observe that we can repeat this argument with $B_{4\delta}(x_k)$ in place of B_1 to get a new covering $\{B_{(4\delta)^2}(x_{k,l})\}_{l=1,2,\dots,N_k}$ of $E \cap B_{4\delta}(x_k)$ with $N_k(4\delta)^{m+\beta} < 1/2$. Thus $\{B_{(4\delta)^2}(x_{k,l})\}_{k=1,2,\dots,N, l=1,2,\dots,N_k}$ covers E with $x_{k,l} \in E$ and $\sum_{k=1}^N N_k(4\delta)^{2 \cdot (m+\beta)} < (1/2)^2$. Repeating this argument for a total of j times, we get a finite covering of E by M balls with centers on E, radii $= (4\delta)^j$, and $M(4\delta)^{j(m+\beta)} < (1/2)^j$. Thus $\mathcal{H}^{m+\beta}(E) \leq C(1/2)^j$ for every integer $j = 1, 2, 3, \dots$. Letting $j \to \infty$, we get $\mathcal{H}^{m+\beta}(E) = 0$.

Proof (*Proof of Proposition 12*) It follows from Lemma 9 and the definition of Hausdorff dimension.

Acknowledgments The author was supported by the European Research Council under the Grant Agreement No. 801867 "Regularity and singularities in elliptic PDE (EllipticPDE)", by the Swiss National Science Foundation, and by MINECO grant MTM2017-84214-C2-1-P.

These lecture notes are based on Chapter 5 of the forthcoming book *Regularity theory for elliptic PDE*, X. Fernández-Real, X. Ros-Oton (2019). I would like to thank X. Fernández-Real and J. Serra for their comments and suggestions on a preliminary version of these notes.

References

1. M. Allen, H. Shahgholian, A new boundary Harnack principle (equations with right hand side). Arch. Rat. Mech. Anal. **234**, 1413–1444 (2019)
2. I. Athanasopoulos, L. Caffarelli, Optimal regularity of lower dimensional obstacle problems. Zap. Nauchn. Sem. S.-Peterburg. Otdel. Mat. Inst. Steklov. (POMI) **310** (2004), Kraev. Zadachi Mat. Fiz. i Smezh. Vopr. Geor. Funkts. **34**, 49–66, 226; reprinted in J. Math. Sci. (N.Y.) **132**, 274–284 (2006)
3. I. Athanasopoulos, L. Caffarelli, S. Salsa, The structure of the free boundary for lower dimensional obstacle problems. Am. J. Math. **130**, 485–498 (2008)
4. B. Barrios, A. Figalli, X. Ros-Oton, Global regularity for the free boundary in the obstacle problem for the fractional Laplacian. Am. J. Math. **140**, 415–447 (2018)
5. L. Caffarelli, Further regularity for the Signorini problem. Commun. Partial Differ. Equ. **4**, 1067–1075 (1979)
6. L. Caffarelli, The obstacle problem revisited. J. Fourier Anal. Appl. **4**, 383–402 (1998)
7. L. Caffarelli, S. Salsa, *A Geometric Approach to Free Boundary Problems* (American Mathematical Society, Providence, 2005)
8. M. Colombo, L. Spolaor, B. Velichkov, Direct epiperimetric inequalities for the thin obstacle problem and applications. Commun. Pure Appl. Math. **72**, 384–420 (2020)
9. R. Cont, P. Tankov, *Financial Modeling with Jump Processes*. Chapman & Hall/CRC Financial Mathematics Series (Chapman & Hall/CRC, Boca Raton, 2004)
10. D. Danielli, N. Garofalo, A. Petrosyan, T. To, Optimal regularity and the free boundary in the parabolic Signorini problem. Mem. Am. Math. Soc. **249**(1181), v + 103pp (2017)
11. E. De Giorgi, Sulla differenziabilità e l'analiticità delle estremali degli integrali multipli regolari. Memorie della Accademia delle Scienze di Torino. Classe di Scienze Fisiche, Matematicahe e Naturali **3**, 25–43 (1957)
12. D. De Silva, O. Savin, A note on higher regularity boundary Harnack inequality. Disc. Cont. Dyn. Syst. **35**, 6155–6163 (2015)
13. D. De Silva, O. Savin, Boundary Harnack estimates in slit domains and applications to thin free boundary problems. Rev. Mat. Iberoam. **32**, 891–912 (2016)
14. G. Duvaut, J.L. Lions, *Inequalities in Mechanics and Physics*. Grundlehren der Mathematischen Wissenschaften, vol. 219 (Springer, Berlin/Heidelberg/New York, 1976)
15. X. Fernández-Real, Y. Jhaveri, On the singular set in the thin obstacle problem: higher order blow-ups and the very thin obstacle problem (Anal. PDE, 2020, in press)
16. X. Fernández-Real, X. Ros-Oton, *Regularity Theory for Elliptic PDE*, forthcoming book, submitted (2019)
17. A. Figalli, J. Serra, On the fine structure of the free boundary for the classical obstacle problem. Invent. Math. **215**, 311–366 (2019)
18. A. Figalli, X. Ros-Oton, J. Serra, Generic regularity of free boundaries for the obstacle problem. Publ. Math. IHÉS **132**, 181–292 (2020)
19. M. Focardi, E. Spadaro, On the measure and the structure of the free boundary of the lower dimensional obstacle problem. Arch. Rat. Mech. Anal. **230**, 125–184 (2018)
20. A. Friedman, *Variational Principles and Free Boundary Problems* (Dover, New York, 1988)

21. N. Garofalo, A. Petrosyan, Some new monotonicity formulas and the singular set in the lower dimensional obstacle problem. Invent. Math. **177**, 414–461 (2009)
22. D. Kinderlehrer, G. Stampacchia, *An Introduction to Variational Inequalities and Their Applications* (SIAM, New York, 1980)
23. H. Koch, A. Petrosyan, W. Shi, Higher regularity of the free boundary in the elliptic Signorini problem. Nonlinear Anal. **126**, 3–44 (2015)
24. B. Krummel, N. Wickramasekera, Fine properties of branch point singularities: two-valued harmonic functions. Preprint arXiv (2013)
25. R. Merton, Option pricing when the underlying stock returns are discontinuous. J. Finan. Econ. **5**, 125–144 (1976)
26. J.F. Nash, Parabolic equations. Proc. Nat. Acad. Sci. U. S. A. **43**, 754–758 (1957)
27. J.F. Nash, Continuity of solutions of parabolic and elliptic equations. Am. J. Math. **80**, 931–954 (1958)
28. A. Petrosyan, H. Shahgholian, N. Uraltseva, *Regularity of Free Boundaries in Obstacle-Type Problems*. Graduate Studies in Mathematics, vol. 136 (American Mathematical Society, Providence, 2012)
29. J.-F. Rodrigues, *Obstacle Problems in Mathematical Physics*. North-Holland Mathematics Studies vol. 134 (North-Holland, Amsterdam/New York, 1987)
30. X. Ros-Oton, Obstacle problems and free boundaries: an overview. SeMA J. **75**, 399–419 (2018)
31. X. Ros-Oton, D. Torres-Latorre, New boundary Harnack inequalities with a right hand side. Preprint arXiv (2020)
32. D.G. Schaeffer, Some examples of singularities in a free boundary. Ann. Scuola Norm. Sup. Pisa **4**, 133–144 (1977)
33. W. Shi, An epiperimetric inequality approach to the parabolic Signorini problem. Disc. Cont. Dyn. Syst. A **40**, 1813–1846 (2020)
34. A. Signorini, Sopra alcune questioni di elastostatica. Atti Soc. It. Progr. Sc. **21**(2), 143–148 (1933)
35. A. Signorini, Questioni di elasticità non linearizzata e semilinearizzata, Rend. Mat. e Appl. **18**, 95–139 (1959)
36. G.S. Weiss, A homogeneity improvement approach to the obstacle problem. Invent. Math. **138**, 23–50 (1999)
37. B. White, Stratification of minimal surfaces, mean curvature flows, and harmonic maps. J. Reine Angew. Math. **488**, 1–35 (1997)

Chapter 4
Bernoulli Type Free Boundary Problems and Water Waves

Georg S. Weiss

Abstract Free boundaries are—a priori not fixed and possibly moving—interfaces separating different phases, e.g. the surfaces of water waves or ice crystals. These objects have interested humans even in very early cultures. The mathematical analysis of free boundaries goes back as far as to Newton who was interested in the science of water waves.

In this course I will introduce Bernoulli type free boundary problems which play an important role in free surface fluid flow. Apart from basic regularity results I will discuss singularities as well as free boundary theory of water waves.

4.1 Introduction

A "free boundary problem" is a problem, where one solves a given partial differential equation in a domain that is not known a priori, meaning that there is no way to know the domain *before* knowing the solution. The boundary of such a domain is called "free boundary" (see Fig. 4.1). To avoid an underdetermined problem-setting, it is necessary to add an additional boundary condition on the free boundary to the usual boundary condition. As the free boundary depends on the solution, the problem as a whole is going to be a nonlinear problem even if one starts with a linear differential equation in each phase.

For motivation let us consider the process of melting ice into water. Unless one does an experiment, one has in this case no means of knowing the position of the boundary between ice and water. Moreover the boundary is moving in time. We may assume that the temperature is in each phase (solid and liquid) a solution of the heat equation, but finding an adequate condition on the free boundary separating the two phases is not an easy task. If one assumes for example that the free boundary is the 0-level set of the temperature, one is thereby excluding phenomena like

G. S. Weiss (✉)
Faculty of Mathematics, University of Duisburg-Essen, Duisburg, Germany
e-mail: georg.weiss@uni-due.de

© The Author(s), under exclusive license to Springer Nature Switzerland AG 2021
M. Focardi, E. Spadaro (eds.), *Geometric Measure Theory and Free Boundary Problems*, Lecture Notes in Mathematics 2284,
https://doi.org/10.1007/978-3-030-65799-4_4

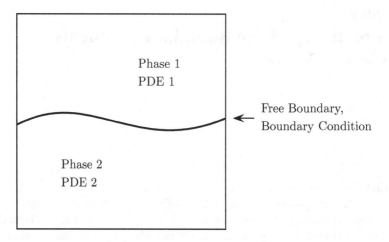

Fig. 4.1 Free Boundary Problem

undercooling. In the case of the Stefan problem with Gibbs-Thomson law ([27]), it is possible to model undercooling, but the problem becomes harder to handle from the mathematical point of view since the boundary between ice and water is not a level set any more.

The industrial applications of free boundary problems are so numerous that a mere overview would be beyond the scope of this editorial. Here we confine ourselves to the explanation of a few examples in some detail. Solidification and melting processes as well as phase separation processes are often mentioned as classical applications of free boundary problems, but there are many other applications as various as fluid jets, tumor growth, Black-Scholes market models and others.

Let us mention that from the point of view of the engineer, classical solutions of free boundary problems—meaning solutions with free boundaries that are locally smooth surfaces and can be extended to smooth functions on the closure of their respective phases—are to be desired. Reasons for this desire are that numerical analysis and the compactness of free boundaries which is necessary in inverse problems become much easier in the case of classical solutions. From the theoretical point of view however the situation is different: the existence of classical solutions of free boundary problems can rarely be shown in a direct way. If all goes well, one obtains a weak solution (e.g. a solution in the sense of distributions or a viscosity solution) and one has a priori no information on the regularity of that solution or the regularity of the free boundary. In the case of the classical Stefan problem, for example, a priori one does not even know whether the weak solution is continuous and whether the free boundary is a relatively closed set. Even when one has proved that the weak solution is continuous, the boundaries of the phases, $\partial\{u > 0\}$ and $\partial\{u < 0\}$ may have positive volume (think for example of a generalized Cantor set). Although the free boundary may be a level set, it is usually not possible to apply undergraduate tools like the implicit function theorem directly, as the solution or derivatives are discontinuous. In the example of the classical Stefan problem the first derivatives are discontinuous.

In these circumstances, regularity theory plays a vital role. The regularity theory of free boundary problems is—apart from the obvious relation to the regularity theory of nonlinear partial differential equations—deeply related to geometric measure theory and harmonic analysis. Much progress has been achieved here during the last few decades, the starting point being the "flatness implies regularity" results that have been proved in the time from the seventies to the early nineties. Roughly speaking, "flatness implies regularity" means that if the free boundary is in the neighborhood of a certain point sufficiently close to a plane, then it has to be a smooth surface in a certain smaller neighborhood of that point. Based on this success in proving regularity in the neighborhood of points that are in the above sense "flat", research is now focusing on the study of singular points (free boundary points that are not flat), free boundaries moving in time, and systems.

In what follows we will discuss the analysis of the free surface (in particular singularities) of water waves from the perspective of Bernoulli type free boundary problems.

4.2 Relation to Water Waves

Water waves have been intriguing objects to men since the early cultures. Isaac Newton was the first scientist to develop a theory of water waves. After the introduction of the equations of hydrodynamics by L. Euler, quite a few famous mathematicians worked on water waves, including P.S. Laplace, A.-L. Cauchy, S. D. Poisson and G. G. Stokes. From a historical point of view the different approaches and the antagonism between the French and the English school in the nineteenth century make for an excellent reading. Let us shortly introduce our model and describe known results.

Consider a wave of permanent form moving with constant speed on the free surface of an incompressible inviscid fluid, acted on by gravity.

With respect to a frame of reference moving with the speed of the wave, the flow is steady and occupies a fixed region D in the plane. The boundary ∂D of the fluid region contains a part $\partial_a D$ which is free and in contact with an air region. Under the assumption that the fluid region D is simply connected, the incompressibility condition shows that the flow can be described by a *stream function* $\psi : D \to \mathbf{R}$, so that the relative fluid velocity is $(\psi_y, -\psi_x)$. The Euler equations imply that the *vorticity* $\omega := -\Delta\psi$ satisfies (Fig. 4.2)

$$\omega_x \psi_y = \omega_y \psi_x \quad \text{in } D. \tag{4.1}$$

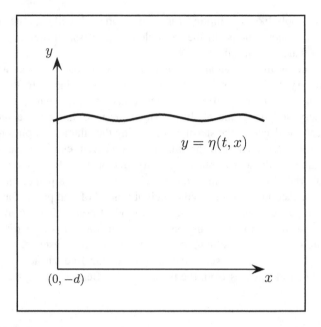

Fig. 4.2 Water wave

It is easy to see that (4.1) is satisfied whenever

$$\omega = \gamma(\psi) \quad \text{in } D \tag{4.2}$$

for some (smooth) function γ of variable ψ, which will be referred to as a *vorticity function*. (Conversely, under additional assumptions, see [11], (4.1) implies the existence of such a function γ.) The kinematic boundary condition that the same particles always form the free surface $\partial_a D$ is equivalent to

$$\psi \text{ is locally constant on } \partial_a D.$$

Also, in the presence of (4.2), Bernoulli's Theorem and the fact that on the fluid-air interface $\partial_a D$ the pressure in the fluid equals the constant atmospheric pressure imply that

$$\frac{1}{2}|\nabla\psi|^2 + gy \text{ is locally constant on } \partial_a D, \tag{4.3}$$

where $g > 0$ is the gravitational constant. We therefore obtain, after some normalization, and at least in the case when $\partial_a D$ is connected, that the following equations and boundary conditions are satisfied:

$$-\Delta\psi = \gamma(\psi) \quad \text{in } D, \tag{4.4a}$$

$$\psi = 0 \quad \text{on } \partial_a D, \tag{4.4b}$$

$$|\nabla\psi|^2 + 2gy = 0 \quad \text{on } \partial_a D. \tag{4.4c}$$

Equations (4.4) are usually supplemented by suitable boundary conditions on the fixed boundary portion of D, or some conditions on the flow at infinity if the fluid domain is unbounded. Classical types of waves which have received most attention in the literature are periodic and solitary waves of finite depth (in which the fluid domain D has a fixed flat bottom $y = -d$, at which ψ is constant), as well as periodic waves of infinite depth (in which the fluid domain extends to $y = -\infty$ and the condition $\lim_{y \to -\infty} \nabla\psi(x, y) = (0, -c)$ holds, where c is the speed of the wave). Conversely, for any vorticity function γ, any solution of (4.4) gives rise to a traveling free-surface gravity water wave, irrespective of whether D is simply connected or $\partial_a D$ is connected. The above problem describes apart from water waves the equally physical problem of the equilibrium state of a fluid when pumping in water from one lateral boundary and sucking it out at the other lateral boundary. In the latter setting we would consider a finite domain with an inhomogeneous Neumann boundary condition at the lateral boundary, and the bottom could be a non-flat surface. Note however that concerning existence there is a huge difference between the two problems.

Here we are interested particularly in the shape of the free boundary $\partial\{\psi > 0\}$ at *stagnation points*, which are points where the relative fluid velocity $(\psi_y, -\psi_x)$ is the zero vector. The Bernoulli condition (4.3) shows that such points are on the real axis, while the rest of the free boundary is in the lower half-plane. Stokes [34] conjectured that, in the irrotational case $\gamma \equiv 0$, at any stagnation point the free surface has a (symmetric) corner of $120°$, and formal asymptotics suggest that the same result might be true also in the general case of waves with vorticity $\gamma \not\equiv 0$.

Stokes gave a formal argument in support of his conjecture, which can be found at the end of this introduction, but a rigorous proof has not been given until 1982, when Amick, Fraenkel and Toland [5] and Plotnikov [29] proved the conjecture independently in brilliant papers. These proofs use an equivalent formulation of problem as a nonlinear singular integral equation due to Nekrasov (derived via conformal mapping), and are based on rather formidable estimates for this equation. In addition, Plotnikov's proof uses ordinary differential equations in the complex plane. Moreover Plotnikov and Toland proved convexity of the two branches of the free surface [30]. A geometric proof not relying on the isolatedness, symmetry, and monotonicity assumptions present in both [5] and [29] has recently been given in [37].

Prior to these works on the Stokes conjecture, the existence of extreme periodic waves, of finite and infinite depth, has been established by Toland [35] and McLeod [26], building on earlier existence results for large-amplitude smooth waves by Krasovskii [25] and by Keady and Norbury [23]. The existence of large-amplitude smooth solitary waves and of extreme solitary waves has then been shown by Amick and Toland [6].

In this course we will focus on the analysis of stagnation points of steady two-dimensional gravity water waves with vorticity in the absence of structural assumptions of isolatedness of stagnation points, symmetry and monotonicity of the free boundary, which have been essential assumptions in all previous results. We will show for example that, in the case when the free surface is an injective curve, the asymptotics at any stagnation point is given either by the "Stokes corner flow" where the free surface has a *corner of* 120° (see Fig. 4.3), or the free surface ends in a *horizontal cusp*, or the free surface is *horizontally flat* at the stagnation point (Fig. 4.4).

The cusp case is a new feature in the case with vorticity, and it is not possible without the presence of vorticity [37] (Fig. 4.5). We conjecture that cusps—the

Fig. 4.3 Stokes corner

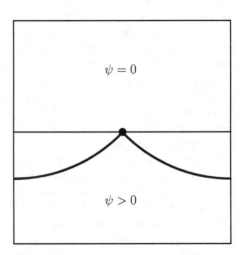

Fig. 4.4 Horizontally flat stagnation point

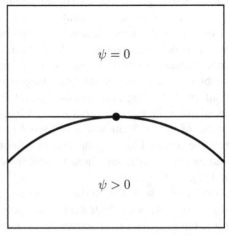

Fig. 4.5 Cusp

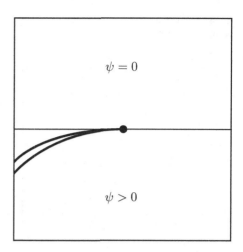

existence of which is still open—are due to the break-down of the Rayleigh-Taylor condition in the presence of vorticity.

4.3 Background Knowledge on the Bernoulli Problem

In the celebrated paper [2], H.W. Alt and L.A. Caffarelli proved via minimization of the energy $\int (|\nabla u|^2 + \chi_{\{u>0\}})$—here $\chi_{\{u>0\}}$ denotes the characteristic function of the set $\{u > 0\}$—existence of a stationary solution of (4.6) in the sense of distributions. They also derived regularity of the free boundary $\partial\{u > 0\}$ up to a set of vanishing $n - 1$-dimensional Hausdorff measure. The question of the existence of classical solutions is by [40] related to non-existence of singular minimizing cones. *Non-minimizing* singular cones *do* in fact appear for $n = 3$ (see [2, example 2.7]). In three dimensions the question of non-existence of singular minimizing cones has been settled in an affirmative way [10], thereby also proving existence of smooth solutions (Fig. 4.6). Figure 4.7 shows an example of a non-minimizing solution with a cusp-shaped free boundary.

It is known, that solutions of the Dirichlet problem (even in two space dimensions) are not unique (see [2, example 2.6]). Still, absolute minimizers of the above energy are ordered so that a comparison principle may be used in some cases. The closest sense of solutions to minimizers are viscosity solutions as viscosity solutions do possess an inherit stability. Actually there are several notions of viscosity solutions but we will not discuss those in detail. Both absolute minimizers and viscosity solutions are not suitable for the application to water waves as they would choose the trivial flat wave. Critical points of the energy are of course relevant to water waves, however we have to discuss which variation of the energy we are talking about as the energy is not differentiable in every direction. One possible choice is the domain variation/inner variation. Those solutions will be called variational

Fig. 4.6 Free boundary in
the Bernoulli problem

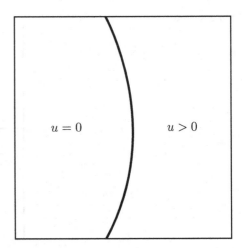

Fig. 4.7 Cusps do occur for
non-minimizing solutions

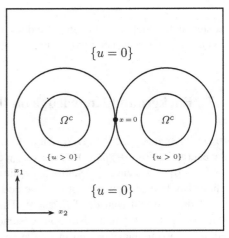

solutions in the sequel and will be defined in detail in Sect. 4.5. Of course variational
solutions do not contain the full information, so the domain variation equation has to
be supplemented by additional information obtained for example from the existence
proof.

Let us also shortly mention the standard approach to regularity in the one-phase
Bernoulli problem: Similarly to the obstacle problem, the standard approach relies
on the following steps:

- Step 1: Derive some form of first variation of the energy.
- Step 2: Prove optimal regularity of the solution (in the case of minimizers,
  Lipschitz regularity). Note that for general weak solutions, optimal regularity
  is not well understood at all.
- Step 3: Non-Degeneracy of minimizers (for general weak solution you may have
  to assume some form of Non-Degeneracy).

Fig. 4.8 Flatness

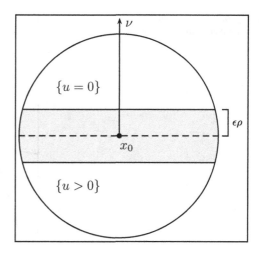

- Step 4: Estimates on the density of the two phases (for minimizers).
- Final Aim is a flatness-implies-regularity result: If the free boundary is flat on the side of $\{u = 0\}$, then in a smaller neighborhood it is a $C^{1,\alpha}$-surface.
  Flatness on the side of $\{u = 0\}$ means that (cf. Fig. 4.8),

$$u = 0 \text{ in } B_r(x_0) \cap \{(x - x_0) \cdot \nu \geq \epsilon r\}.$$

Here x_0 is a free boundary point, C and c are constants and ν is a unit vector.

The proof of the flatness-implies-regularity result, however, relies on different methods from those known for the obstacle problem. In [2], crucial ideas include: 1. A "partition of energy'. "Partition of energy" means here that the Dirichlet part of the energy is asymptotically less than or equal to the volume part of the energy, i.e.

$$r^{-n} \int_{B_r(x_0)} |\nabla u|^2 \leq C r^\alpha + r^{-n} \int_{B_r(x_0)} \chi_{\{u > 0\}} \; ;$$

here, too, x_0 is assumed to be a free boundary point. In two space dimensions the inequality becomes an equality, that is, an "equipartition of energy". "Equipartition of energy" with respect to a different energy functional has later played a vital role in the context of mean curvature flow.

2. Non-positive mean curvature of the free boundary of blow-up limit. "Non-positive mean curvature of the free boundary of blow-up limits" refers to the fact that each limit u_0 of the blow-up sequence $(\frac{u(x_m + r_m \cdot)}{r_m})_{m \in \mathbf{N}}$ satisfies for every open test set D the inequality $\mathcal{H}^{n-1}(D \cap \partial\{u_0 > 0\}) \leq \mathcal{H}^{n-1}(\partial D \cap \{u_0 > 0\})$.

Let us point out, however, that the idea behind this fact has in [2] been applied to inhomogeneous blow-up limits and not to the homogeneous blow-up limits from above. Better regularity for inhomogeneous blow-up limits is used in [2] to derive

Fig. 4.9 Idea behind the
non-positive mean curvature
property

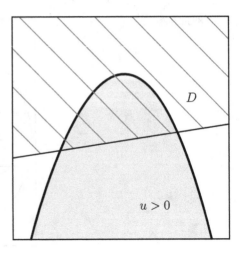

a quantitative estimate, which eventually leads to regularity of the free boundary
(Fig. 4.9).

While these ideas are really sophisticated, they are in a way too refined to be
extended to more difficult settings. Actually (cf. [7]) a linearization in a cruder
scaling leads more easily to the condition on the free boundary in a linearization.
Interestingly, in the two-phase problem, the linearization may lead to a *nonlinear
problem*, in this case the thin obstacle problem (see [8])!

Concerning the Bernoulli problem, the singular perturbation approach

$$\Delta u_\epsilon = \beta_\epsilon(u_\epsilon) \text{ in } \Omega \tag{4.5}$$

has been very popular. Here $\epsilon \in (0, 1)$, $\beta_\epsilon(z) = \frac{1}{\epsilon}\beta(\frac{z}{\epsilon})$, $\beta \in C_0^1([0, 1])$, $\beta > 0$
in $(0, 1)$ and $\int \beta = \frac{1}{2}$. Formally, each limit u with respect to a sequence $\epsilon_m \to 0$ is
a solution of the free boundary problem

$$\Delta u = 0 \text{ in } \{u > 0\} \cap \Omega, \ |\nabla u| = 1 \text{ on } \partial\{u > 0\} \cap \Omega . \tag{4.6}$$

Let us point out that one cannot hope the limit to be a distributional solution in the
sense that

$$\int_0^\infty \int_{\mathbf{R}^n} u \Delta\phi = \int_{\partial\{u>0\}} \phi \, d\mathcal{H}^n \text{ for } \phi \in C_0^\infty(\mathbf{R}^n) \tag{4.7}$$

(here $\mathcal{H}^n$ denotes the n-dimensional Hausdorff measure): solving the one-
dimensional stationary equation (4.5), more precisely,

$$u_\epsilon'' = \beta_\epsilon(u_\epsilon) \text{ in } \mathbf{R}, \ u_\epsilon'(0) = 0, \ u_\epsilon(0) = \begin{cases} \mathcal{B}_\epsilon^{-1}(\delta) \text{ for } \delta \in (0, \frac{1}{2}] \\ \mathcal{B}_\epsilon^{-1}(\epsilon) \text{ for } \delta = 0 \end{cases} \tag{4.8}$$

(here $\mathcal{B}(z) := \int_0^z \beta(s)\, ds$, $\mathcal{B}_\epsilon(z) := \int_0^z \beta_\epsilon(s)\, ds$ and $\mathcal{B}_\epsilon^{-1}$ is the inverse of the restriction of $\mathcal{B}_\epsilon$ to the interval $[0, \epsilon]$), the solution $u_\epsilon(x)$ converges to $\sqrt{1 - 2\delta}|x|$ as $\epsilon \to 0$. The function $|x_n|$ also appears as the blow-up limit of the stationary *distributional* solution of (4.6), $u(x) = \max(-\log(|x - e_1|), 0) + \max(-\log(|x + e_1|), 0)$ in $\mathbf{R}^2 - (B_{\frac{1}{2}}(e_1) \cup B_{\frac{1}{2}}(-e_1))$, at the point $x = 0$ (see Fig. 4.7). It also appears as the blow-up limit (with respect to the parabolic metric) of time-dependent examples at the first-in-time singularity of the free boundary.

Thus another reason for not confining ourselves to solutions in the sense of (4.7) is to create a class of solutions that is *closed with respect to the blow-up process*.
In [42] we considered the time-dependent (parabolic) version of (4.5). As an intermediate result we obtained that each limit u of the singular perturbation problem

$$\partial_t u_\epsilon - \Delta u_\epsilon = -\beta_\epsilon(u_\epsilon) \text{ in } (0, \infty) \times \mathbf{R}^n$$

is a solution *in the sense of domain variations*, i.e. u is smooth in $\{u > 0\}$ and satisfies

$$\int_0^\infty \int_{\mathbf{R}^n} [-2\partial_t u \nabla u \cdot \xi + |\nabla u|^2 \text{div}\, \xi - 2\nabla u D\xi \nabla u] = -\int_0^\infty \int_{R(t)} \xi \cdot v\, d\mathcal{H}^{n-1}\, dt$$

(4.9)

for every $\xi \in C_0^{0,1}((0, \infty) \times \mathbf{R}^n; \mathbf{R}^n)$. Here

$$R(t) := \{x \in \partial\{u(t) > 0\} \ : \ \text{there is } v(t, x) \in \partial B_1(0) \text{ such that } u_r(s, y) =$$

$$\frac{u(t + r^2 s, x + ry)}{r} \to \max(-y \cdot v(t, x), 0) \text{ locally uniformly in } (s, y) \in \mathbf{R}^{n+1}$$

$$\text{as } r \to 0\}$$

is a countably $n - 1$-rectifiable subset of the free boundary. The set $R(t)$ is "almost" open relative to $\partial\{u(t) > 0\}$ in some sense. Let us remark that already the above equation contains information (apart from the rectfiability and almost openness of $R(t)$) that cannot be inferred from viscosity notions.

In addition we proved that each limit of the time-dependent version of (4.5)—no additional assumptions are necessary—satisfies for a.e. $t \in (0, \infty)$ the following:

$$\int_{\mathbf{R}^n} (\partial_t u(t)\phi + \nabla u(t) \cdot \nabla\phi) = -\int_{R(t)} \phi\, d\mathcal{H}^{n-1} - \int_{\Sigma_{**}(t)} 2\theta(t, \cdot)\, \phi\, d\mathcal{H}^{n-1}$$
$$- \int_{\Sigma_z(t)} \phi\, d\lambda(t)$$

(4.10)

for every $\phi \in C_0^{0,1}(\mathbf{R}^n)$, the graph of $u(t)$ has a *unique tangent cone* at $\mathcal{H}^{n-1}$-a.e. point, and the free boundary $\partial\{u(t) > 0\}$ can up to a set of vanishing $\mathcal{H}^{n-1}$ measure be decomposed as $\partial\{u(t) > 0\} = R(t) \cup \Sigma_{**}(t) \cup \Sigma_z(t)$. Here the non-degenerate singular set

$$\Sigma_{**}(t) := \{x \in \partial\{u(t) > 0\} : \text{ there is } \theta(t, x) \in (0, 1] \text{ and } \xi(t, x) \in \partial B_1(0) \text{ such}$$

that $u_r(s, y) = \dfrac{u(t + r^2 s, x + ry)}{r} \to \theta(t, x)|y \cdot \xi(t, x)|$ locally uniformly

$$\text{in } (s, y) \in \mathbf{R}^{n+1} \text{ as } r \to 0\}$$

is a countably $n - 1$-rectifiable subset of the free boundary, $\lambda(t)$ is a Radon measure satisfying $r^{1-n}\lambda(t)(B_r(x)) \to 0$ for $\mathcal{H}^{n-1}$-a.e. $x \in \mathbf{R}^n$, and the support of $\lambda(t)$ is contained in the degenerate singular set

$$\Sigma_z(t) := \{x \in \partial\{u(t) > 0\} : r^{-n-2} \int_{Q_r(t,x)} |\nabla u|^2 \to 0 \text{ as } r \to 0\} .$$

The decomposition says therefore that the free boundary consists of a countably $n - 1$-rectifiable part and the degenerate singular set. An assumption like $\mathcal{H}^{n-1}(\Sigma_z(t)) < \infty$ would incidentally imply that $\lambda(t) = 0$.

Let us remark that our result leads to the interesting questions, whether solutions with $\lambda \neq 0$ exist and whether the set of free boundary points (t_0, x_0) such that $r^{-2-n} \int_{Q_r(t_0,x_0)} |\nabla u|^2 \to 0$ as $r \to 0$ can possibly form a large set (an example for the existence of such points is the solution $u(x) = |x_1^2 - x_2^2|$ in $\mathbf{R}^2$). Concerning the first question in the two-dimensional stationary case, D. Jerison and the author obtained from [44] that $\lambda = 0$, i.e. the term does not appear in the equation. Concerning the first question in higher dimensions and the second question, examples of extremely irregular harmonic measures like [45, Example 3] as well as the fact, that there exists a sequence of stationary solutions of (4.6),

$$u_m(x) = \frac{1}{m} \max(\log(m|x|), 0) \text{ in } \mathbf{R}^2 , \qquad (4.11)$$

satisfying $\int_{B_1(0)} |\nabla u_m|^2 \to 0$ and $\sup_{B_1(0)} |\nabla u_m| = 1$ as $m \to \infty$, do not allow much hope to exclude the possibility of nontrivial λ by a general principle.

To conclude the present Section, let us mention that although the successful techniques in the Bernoulli problem give rise to hope for an alternative existence proof for water waves, due to lack of compactness and the crucial problem of avoiding the trivial flat wave, conformal maps (mostly combined with bifurcation theory) have been the only successful approach to the existence of water waves to this date.

4.4 Notation

We denote by χ_A the characteristic function of a set A. For any real number a, the notation a^+ stands for $\max(a, 0)$. We denote by $x \cdot y$ the Euclidean inner product in $\mathbf{R}^n \times \mathbf{R}^n$, by $|x|$ the Euclidean norm in $\mathbf{R}^n$ and by $B_r(x^0) := \{x \in \mathbf{R}^n : |x - x^0| < r\}$ the ball of center x^0 and radius r. We will use the notation B_r for $B_r(0)$, and denote by ω_n the n-dimensional volume of B_1. Also, $\mathcal{L}^n$ shall denote the n-dimensional Lebesgue measure and $\mathcal{H}^s$ the s-dimensional Hausdorff measure. By ν we will always refer to the outer normal on a given surface. We will use functions of bounded variation $BV(U)$, i.e. functions $f \in L^1(U)$ for which the distributional derivative is a vector-valued Radon measure. Here $|\nabla f|$ denotes the total variation measure (cf. [21]). Note that for a smooth open set $E \subset \mathbf{R}^n$, $|\nabla \chi_E|$ coincides with the surface measure on ∂E.

The subsequent sections follow closely the result [38].

4.5 Notion of Solution

In some sections of the paper we work with a n-dimensional generalization of the problem described in Sect. 4.2. Let Ω be a bounded domain in $\mathbf{R}^n$ which has a non-empty intersection with the hyperplane $\{x_n = 0\}$, in which to consider the combined problem for fluid and air. We study solutions u, in a sense to be specified, of the problem

$$\Delta u = -f(u) \quad \text{in } \Omega \cap \{u > 0\}, \tag{4.12}$$

$$|\nabla u|^2 = x_n \quad \text{on } \Omega \cap \partial\{u > 0\}.$$

Note that, compared to Sect. 4.2, we have switched notation from ψ to u and from γ to f, and we have "reflected" the problem at the hyperplane $\{x_n = 0\}$. The nonlinearity f is assumed to be a continuous function with primitive $F(z) = \int_0^z f(t) \, dt$. Since our results are completely local, we do not specify boundary conditions on $\partial\Omega$. In view of the second equation in (4.12), it is natural to assume throughout the rest of the paper that $u \equiv 0$ in $\Omega \cap \{x_n \leq 0\}$.

We begin by introducing our notion of a *variational solution* of problem (4.12).

Definition 1 (Variational Solution) We define $u \in W^{1,2}_{\text{loc}}(\Omega)$ to be a *variational solution* of (4.12) if $u \in C^0(\Omega) \cap C^2(\Omega \cap \{u > 0\})$, $u \geq 0$ in Ω and $u \equiv 0$ in $\Omega \cap \{x_n \leq 0\}$, and the first variation with respect to domain variations of the functional

$$J(v) := \int_\Omega \left(|\nabla v|^2 - 2F(v) + x_n \chi_{\{v > 0\}} \right) dx$$

vanishes at $v = u$, i.e.

$$0 = -\frac{d}{d\epsilon} J(u(x + \epsilon\phi(x)))|_{\epsilon=0}$$

$$= \int_\Omega \left((|\nabla u|^2 - 2F(u))\operatorname{div}\phi - 2\nabla u D\phi\nabla u + x_n\chi_{\{u>0\}}\operatorname{div}\phi + \chi_{\{u>0\}}\phi_n \right) dx$$

for any $\phi \in C_0^1(\Omega; \mathbf{R}^n)$.

Note for future reference that for each open set $D \subset\subset \Omega$ there is $C_D < +\infty$ such that $\Delta u + C_D$ is a nonnegative Radon measure in D, the support of the singular part of which (with respect to the Lebesgue measure) is contained in the set $\partial\{u > 0\}$: by Sard's theorem $\{u = \delta\} \cap D$ is for almost every δ a smooth surface. It follows that for every non-negative $\zeta \in C_0^\infty(D)$

$$-\int_D (\nabla\max(u - \delta, 0) \cdot \nabla\zeta - C_D\zeta)\, dx$$

$$= \int_D \zeta(\chi_{\{u>\delta\}}\Delta u + C_D)\, dx - \int_{D\cap\partial\{u>\delta\}} \zeta\nabla u \cdot v\, d\mathcal{H}^{n-1} \geq 0,$$

provided that $|f(u)| \leq C_D$ in D. Letting $\delta \to 0$ and using that u is continuous and nonnegative in Ω, we obtain

$$-\int_D (\nabla u \cdot \nabla\zeta - C_D\zeta)\, dx \leq 0.$$

Thus $\Delta u + C_D$ is a nonnegative distribution in D, and the stated property follows.

Since we want to focus in the present paper on the analysis of stagnation points, we will assume that everything is smooth away from $x_n = 0$, however this assumption may be weakened considerably by using in $\{x_n > 0\}$ regularity theory for the Bernoulli free boundary problem (see [2] for regularity theory in the case $f = 0$—which could effortlessly be perturbed to include the case of bounded f—and see [13] for another regularity approach which already includes the perturbation).

Definition 2 (Weak Solution) We define $u \in W^{1,2}_{loc}(\Omega)$ to be a *weak solution* of (4.12) if the following are satisfied: u is a *variational solution* of (4.12) and the topological free boundary $\partial\{u > 0\} \cap \Omega \cap \{x_n > 0\}$ is locally a $C^{2,\alpha}$-surface.

Remark 1 (i) It follows that in $\{x_n > 0\}$ the solution is a classical solution of (4.12).

(ii) For any weak solution u of (4.12) such that

$$|\nabla u|^2 \leq Cx_n^+ \quad \text{locally in } \Omega,$$

u is a variational solution of (4.12), $\chi_{\{u>0\}}$ is locally in $\{x_n > 0\}$ a function of bounded variation, and the total variation measure $|\nabla \chi_{\{u>0\}}|$ satisfies

$$r^{1/2-n} \int_{B_r(y)} \sqrt{x_n}\, d|\nabla \chi_{\{u>0\}}| \le C_0$$

for all $B_r(y) \subset\subset \Omega$ such that $y_n = 0$ (see [37, Lemma 3.4]).

4.6 Monotonicity Formula and Consequences

A first tool in our analysis is an extension of the monotonicity formula in [41], [40, Theorem 3.1] to the boundary case. The roots of those monotonicity formulas are harmonic mappings [31, 32] and blow-up [28].

Theorem 1 (Monotonicity Formula) *Let u be a variational solution of* (4.12), *let $x^0 \in \Omega$ such that $x_n^0 = 0$, and let $\delta := \mathrm{dist}(x^0, \partial\Omega)/2$. Let, for any $r \in (0, \delta)$,*

$$I_{x^0,u}(r) = I(r) = r^{-n-1} \int_{B_r(x^0)} \left(|\nabla u|^2 - uf(u) + x_n \chi_{\{u>0\}}\right) dx, \tag{4.13}$$

$$J_{x^0,u}(r) = J(r) = r^{-n-2} \int_{\partial B_r(x^0)} u^2 \, d\mathcal{H}^{n-1}, \tag{4.14}$$

$$M_{x^0,u}(r) = M(r) = I(r) - \frac{3}{2} J(r) \tag{4.15}$$

and

$$K_{x^0,u}(r) = K(r) = r \int_{\partial B_r(x^0)} (2F(u) - uf(u)) \, d\mathcal{H}^{n-1}$$

$$+ \int_{B_r(x^0)} ((n-2)uf(u) - 2nF(u)) \, dx. \tag{4.16}$$

Then, for a.e. $r \in (0, \delta)$,

$$I'(r) = r^{-n-2} \left(2r \int_{\partial B_r(x^0)} (\nabla u \cdot v)^2 \, d\mathcal{H}^{n-1} - 3 \int_{\partial B_r(x^0)} u \nabla u \cdot v \, d\mathcal{H}^{n-1}\right)$$

$$+ r^{-n-2} K(r), \tag{4.17}$$

$$J'(r) = r^{-n-3} \left(2r \int_{\partial B_r(x^0)} u \nabla u \cdot v \, d\mathcal{H}^{n-1} - 3 \int_{\partial B_r(x^0)} u^2 \, d\mathcal{H}^{n-1}\right) \tag{4.18}$$

and

$$M'(r) = 2r^{-n-1} \int_{\partial B_r(x^0)} \left(\nabla u \cdot \nu - \frac{3}{2} \frac{u}{r} \right)^2 d\mathcal{H}^{n-1} + r^{-n-2} K(r). \tag{4.19}$$

Proof The identity (4.18) can be easily checked directly, being valid for any function $u \in W^{1,2}_{loc}(\Omega)$ (not necessarily a variational solution of (4.12)).

For small positive κ and $\eta_\kappa(t) := \max(0, \min(1, \frac{r-t}{\kappa}))$, we take after approximation $\phi_\kappa(x) := \eta_\kappa(|x - x^0|)(x - x^0)$ as a test function in the definition of a variational solution. We obtain

$$0 = \int_\Omega \left(|\nabla u|^2 - 2F(u) + x_n \chi_{\{u > 0\}} \right) \left(n \eta_\kappa(|x - x^0|) + \eta'_\kappa(|x - x^0|)|x - x^0| \right) dx$$

$$- 2 \int_\Omega \left(|\nabla u|^2 \eta_\kappa(|x - x^0|) + \nabla u \cdot \frac{x - x^0}{|x - x^0|} \nabla u \cdot \frac{x - x^0}{|x - x^0|} \eta'(|x - x^0|)|x - x^0| \right) dx$$

$$+ \int_\Omega \eta_\kappa(|x - x^0|) x_n \chi_{\{u > 0\}} dx.$$

Passing to the limit as $\kappa \to 0$, we obtain, for a.e. $r \in (0, \delta)$,

$$0 = n \int_{B_r(x^0)} \left(|\nabla u|^2 - 2F(u) + x_n \chi_{\{u > 0\}} \right) dx \tag{4.20}$$

$$- r \int_{\partial B_r(x^0)} \left(|\nabla u|^2 - 2F(u) + x_n \chi_{\{u > 0\}} \right) d\mathcal{H}^{n-1}$$

$$+ 2r \int_{\partial B_r(x^0)} (\nabla u \cdot \nu)^2 d\mathcal{H}^{n-1} - 2 \int_{B_r(x^0)} |\nabla u|^2 dx$$

$$+ \int_{B_r(x^0)} x_n \chi_{\{u > 0\}} dx.$$

Also observe that letting $\epsilon \to 0$ in

$$\int_{B_r(x^0)} \nabla u \cdot \nabla \max(u - \epsilon, 0)^{1+\epsilon} dx = \int_{B_r(x^0)} f(u) \max(u - \epsilon, 0)^{1+\epsilon} dx$$

$$+ \int_{\partial B_r(x^0)} \max(u - \epsilon, 0)^{1+\epsilon} \nabla u \cdot \nu \, d\mathcal{H}^{n-1}$$

for a.e. $r \in (0, \delta)$, we obtain the integration by parts formula

$$\int_{B_r(x^0)} \left(|\nabla u|^2 - uf(u) \right) dx = \int_{\partial B_r(x^0)} u \nabla u \cdot \nu \, d\mathcal{H}^{n-1} \tag{4.21}$$

for a.e. $r \in (0, \delta)$.

Note also that

$$I'(r) = -(n+1)r^{-n-2} \int_{B_r(x^0)} (|\nabla u|^2 - uf(u) + x_n \chi_{\{u>0\}}) \, dx$$

$$+ r^{-n-1} \int_{\partial B_r(x^0)} (|\nabla u|^2 - uf(u) + x_n \chi_{\{u>0\}}) \, d\mathcal{H}^{n-1}. \tag{4.22}$$

Using (4.20) and (4.21) in (4.22), we obtain (4.17). Finally, (4.19) follows immediately by combining (4.17) and (4.18).

From now on we assume

Assumption 2 *Let u satisfy*

$$|\nabla u|^2 \le C x_n^+ \quad \text{locally in } \Omega.$$

Remark 2 Note that Assumption 2 implies that

$$u(x) \le C_1 x_n^{3/2}$$

and that in the case $x_n^0 = 0$,

$$r^{-n-2} |K(r)| \le C_2 \frac{1}{\sqrt{r}},$$

where C_2 depends on x^0 but is locally uniformly bounded.

We first show that the function $M_{x^0,u}$ has a right limit $M_{x^0,u}(0+)$, of which we derive structural properties.

Lemma 1 *Let u be a variational solution of* (4.12) *satisfying Assumption* 2. *Then:*

(i) *Let $x^0 \in \Omega$ be such that $x_n^0 = 0$. Then the limit $M_{x^0,u}(0+)$ exists and is finite. (Note that $u = 0$ in $\{x_n = 0\}$ by assumption.)*

(ii) *Let $x^0 \in \Omega$ be such that $x_n^0 = 0$, and let $0 < r_m \to 0+$ as $m \to \infty$ be a sequence such that the blow-up sequence*

$$u_m(x) := \frac{u(x^0 + r_m x)}{r_m^{3/2}} \tag{4.23}$$

converges weakly in $W_{\text{loc}}^{1,2}(\mathbf{R}^n)$ to a blow-up limit u_0. Then u_0 is a homogeneous function of degree $3/2$, i.e.

$$u_0(\lambda x) = \lambda^{3/2} u_0(x) \quad \text{for any } x \in \mathbf{R}^n \text{ and } \lambda > 0.$$

(iii) *Let u_m be a converging sequence of (ii). Then u_m converges strongly in $W_{\mathrm{loc}}^{1,2}(\mathbf{R}^n)$.*

(iv) *Let $x^0 \in \Omega$ be such that $x_n^0 = 0$. Then*

$$M_{x^0,u}(0+) = \lim_{r \to 0+} r^{-n-1} \int_{B_r(x^0)} x_n^+ \chi_{\{u>0\}} \, dx,$$

and in particular $M_{x^0,u}(0+) \in [0, +\infty)$. Moreover, $M_{x^0,u}(0+) = 0$ implies that $u_0 = 0$ in $\mathbf{R}^n$ for each blow-up limit u_0 of (ii).

(v) *The function $x \mapsto M_{x,u}(0+)$ is upper semicontinuous in $\{x_n = 0\}$.*

(vi) *Let u_m be a sequence of variational solutions of (4.12) with nonlinearity f_m in a domain Ω_m, where*

$$\Omega_1 \subset \Omega_2 \subset \ldots \subset \Omega_m \subset \Omega_{m+1} \subset \ldots \quad \textit{and} \quad \bigcup_{m=1}^{\infty} \Omega_m = \mathbf{R}^n,$$

such that u_m converges strongly to u_0 in $W_{\mathrm{loc}}^{1,2}(\mathbf{R}^n)$, $\chi_{\{u_m>0\}}$ converges weakly in $L_{\mathrm{loc}}^2(\mathbf{R}^n)$ to χ_0, and $f_m(u_m)$ converges to 0 locally uniformly in $\mathbf{R}^n$. Then u_0 is a variational solution of (4.12) with nonlinearity $f = 0$ in $\mathbf{R}^n$ and satisfies the Monotonicity Formula (with $f = 0$), but with $\chi_{\{u_0>0\}}$ replaced by χ_0. Moreover, for each $x^0 \in \mathbf{R}^n$ such that $x_n^0 = 0$, and all instances of $\chi_{\{u_0>0\}}$ replaced by χ_0,

$$M_{x^0,u_0}(0+) \geq \limsup_{m \to \infty} M_{x^0,u_m}(0+).$$

Proof

(i) By Remark 2,

$$u(x) \leq C_2 x_n^{3/2} \quad \text{locally in } \Omega$$

and

$$|r^{-n-2} K(r)| \leq C_3 r^{-1/2} \text{ for each } x^0 \in \Omega \text{ satisfying } x_n^0 = 0. \tag{4.24}$$

Thus $r \mapsto r^{-n-2} K(r)$ is integrable at such points x^0, and from Theorem 1 we infer that the function $M_{x^0,u}$ has a finite right limit $M_{x^0,u}(0+)$.

(ii) For each $0 < \sigma < \infty$ the sequence u_m is by assumption bounded in $C^{0,1}(B_\sigma)$. For any $0 < \varrho < \sigma < \infty$, we write the identity (4.19) in integral form as

$$2 \int_\varrho^\sigma r^{-n-1} \int_{\partial B_r(x^0)} \left(\nabla u \cdot \nu - \frac{3}{2} \frac{u}{r} \right)^2 d\mathcal{H}^{n-1} dr$$

$$= M(\sigma) - M(\varrho) - \int_\varrho^\sigma r^{-n-2} K(r) \, dr. \qquad (4.25)$$

It follows by rescaling in (4.25) that

$$2 \int_{B_\sigma(0) \backslash B_\varrho(0)} |x|^{-n-3} \left(\nabla u_m(x) \cdot x - \frac{3}{2} u_m(x) \right)^2 dx$$

$$\leq M(r_m \sigma) - M(r_m \varrho) + \int_{r_m \varrho}^{r_m \sigma} r^{-n-2} |K(r)| \, dr \to 0 \quad \text{as } m \to \infty,$$

which yields the desired homogeneity of u_0.

(iii) In order to show strong convergence of u_m in $W_{\text{loc}}^{1,2}(\mathbf{R}^n)$, it is sufficient, in view of the weak L^2-convergence of ∇u_m, to show that

$$\limsup_{m \to \infty} \int_{\mathbf{R}^n} |\nabla u_m|^2 \eta \, dx \leq \int_{\mathbf{R}^n} |\nabla u_0|^2 \eta \, dx$$

for each $\eta \in C_0^1(\mathbf{R}^n)$. Let $\delta := \text{dist}(x^0, \partial \Omega)/2$. Then, for each m, u_m is a variational solution of

$$\Delta u_m = -r_m^{1/2} f(r_m^{3/2} u_m) \quad \text{in } B_{\delta/r_m} \cap \{u_m > 0\}, \qquad (4.26)$$

$$|\nabla u_m|^2 = x_n \quad \text{on } B_{\delta/r_m} \cap \partial \{u_m > 0\}.$$

Since u_m converges to u_0 locally uniformly, it follows from (4.26) that u_0 is harmonic in $\{u_0 > 0\}$. Also, using the uniform convergence, the continuity of u_0 and its harmonicity in $\{u_0 > 0\}$ we obtain as in the proof of (4.21) that

$$\int_{\mathbf{R}^n} |\nabla u_m|^2 \eta \, dx = - \int_{\mathbf{R}^n} u_m \left(\nabla u_m \cdot \nabla \eta - r_m^{1/2} f(r_m^{3/2} u_m) \eta \right) dx$$

$$\to - \int_{\mathbf{R}^n} u_0 \nabla u_0 \cdot \nabla \eta \, dx = \int_{\mathbf{R}^n} |\nabla u_0|^2 \eta \, dx$$

as $m \to \infty$. It therefore follows that u_m converges to u_0 strongly in $W_{\text{loc}}^{1,2}(\mathbf{R}^n)$ as $m \to \infty$.

(iv) Let us take a sequence $r_m \to 0+$ such that u_m defined in (4.23) converges weakly in $W_{\text{loc}}^{1,2}(\mathbf{R}^n)$ to a function u_0. Note that by the definition of a variational solution, u_m and u_0 are identically zero in $x_n \le 0$. Using (iii) and the homogeneity of u_0, we obtain that

$$\lim_{m \to \infty} M_{x^0,u}(r_m) = \int_{B_1} |\nabla u_0|^2 \, dx - \frac{3}{2} \int_{\partial B_1} u_0^2 \, d\mathcal{H}^{n-1}$$

$$+ \lim_{r \to 0+} r^{-n-1} \int_{B_r(x^0)} x_n^+ \chi_{\{u>0\}} \, dx$$

$$= \lim_{r \to 0+} r^{-n-1} \int_{B_r(x^0)} x_n^+ \chi_{\{u>0\}} \, dx.$$

Thus $M_{x^0,u}(0+) \ge 0$, and equality implies that for each $\tau > 0$, u_m converges to 0 in measure in the set $\{x_n > \tau\}$ as $m \to \infty$, and consequently $u_0 = 0$ in $\mathbf{R}^n$.

(v) For each $\delta > 0$ we obtain from the Monotonicity Formula (Theorem 1), Remark 2 as well as the fact that $\lim_{x \to x^0} M_{x,u}(r) = M_{x^0,u}(r)$ for $r > 0$, that

$$M_{x,u}(0+) \le M_{x,u}(r) + C\sqrt{r} \le M_{x^0,u}(r) + \frac{\delta}{2} \le M_{x^0,u}(0+) + \delta,$$

if we choose for fixed x^0 first $r > 0$ and then $|x - x^0|$ small enough.

(vi) The fact that u_0 is a variational solution of (4.12) and satisfies the Monotonicity Formula in the sense indicated follows directly from the convergence assumption. The proof for the rest of the claim follows by the same argument as in (v).

4.7   The Two-dimensional Case

Theorem 3 (Two-dimensional Case) *Let $n = 2$, let u be a variational solution of (4.12) satisfying Assumption 2, let $x^0 \in \Omega$ be such that $x_2^0 = 0$, and suppose that*

$$r^{-3/2} \int_{B_r(x^0)} \sqrt{x_2} \, d|\nabla \chi_{\{u>0\}}| \le C_0$$

for all $r > 0$ such that $B_r(x^0) \subset\subset \Omega$. Then the following hold:

(i)

$$M(0+) \in \left\{ 0, \int_{B_1} x_2^+ \chi_{\{x : \pi/6 < \theta < 5\pi/6\}} \, dx, \int_{B_1} x_2^+ \, dx \right\}.$$

(ii) *If $M(0+) = \int_{B_1} x_2^+ \chi_{\{x:\pi/6 < \theta < 5\pi/6\}} dx$, then*

$$\frac{u(x^0 + rx)}{r^{3/2}} \to \frac{\sqrt{2}}{3} \rho^{3/2} \cos(\frac{3}{2}(\min(\max(\theta, \frac{\pi}{6}), \frac{5\pi}{6}) - \frac{\pi}{2})) \quad as\ r \to 0+$$

strongly in $W_{loc}^{1,2}(\mathbf{R}^2)$ and locally uniformly on $\mathbf{R}^2$, where $x = (\rho \cos \theta, \rho \sin \theta)$.

(iii) *If $M(0+) \in \{0, \int_{B_1} x_2^+ dx\}$, then*

$$\frac{u(x^0 + rx)}{r^{3/2}} \to 0 \quad as\ r \to 0+,$$

strongly in $W_{loc}^{1,2}(\mathbf{R}^2)$ and locally uniformly on $\mathbf{R}^2$.

Proof Consider a blow-up sequence u_m as in Lemma 1(ii), where $r_m \to 0+$, with blow-up limit u_0. Because of the strong convergence of u_m to u_0 in $W_{loc}^{1,2}(\mathbf{R}^2)$ and the compact embedding from BV into L^1, u_0 is a homogeneous solution of

$$0 = \int_{\mathbf{R}^2} \left(|\nabla u_0|^2 \mathrm{div}\ \phi - 2\nabla u_0 D\phi \nabla u_0 \right) dx + \int_{\mathbf{R}^2} \left(x_2 \chi_0 \mathrm{div}\ \phi + \chi_0 \phi_2 \right) dx \quad (4.27)$$

for any $\phi \in C_0^1(\mathbf{R}^2; \mathbf{R}^2)$, where χ_0 is the strong L_{loc}^1-limit of $\chi_{\{u_m > 0\}}$ along a subsequence. The values of the function χ_0 are almost everywhere in $\{0, 1\}$, and the locally uniform convergence of u_m to u_0 implies that $\chi_0 = 1$ in $\{u_0 > 0\}$. The homogeneity of u_0 and its harmonicity in $\{u_0 > 0\}$ show that each connected component of $\{u_0 > 0\}$ is a cone with vertex at the origin and of opening angle $120°$. Since $u = 0$ in $\{x_2 \leq 0\}$, this shows that $\{u_0 > 0\}$ has at most one connected component. Note also that (4.27) implies that χ_0 is constant in each open connected set $G \subset \{u_0 = 0\}^\circ$ that does not intersect $\{x_2 = 0\}$.

Consider first the case when $\{u_0 > 0\}$ is non-empty, and is therefore a cone as described above. Let z be an arbitrary point in $\partial \{u_0 > 0\} \setminus \{0\}$. Note that the normal to $\partial \{u_0 > 0\}$ has the constant value $v(z)$ in $B_\delta(z) \cap \partial \{u_0 > 0\}$ for some $\delta > 0$. Plugging in $\phi(x) := \eta(x)v(z)$ into (4.27), where $\eta \in C_0^1(B_\delta(z))$ is arbitrary, and integrating by parts, it follows that

$$0 = \int_{\partial \{u_0 > 0\}} \left(-|\nabla u_0|^2 + x_2(1 - \bar{\chi}_0) \right) \eta\, d\mathcal{H}^1. \quad (4.28)$$

Here $\bar{\chi}_0$ denotes the constant value of χ_0 in the respective connected component of $\{u_0 = 0\}^\circ \cap \{x_2 \neq 0\}$. Note that by Hopf's principle, $\nabla u_0 \cdot v \neq 0$ on $B_\delta(z) \cap \partial \{u_0 > 0\}$. It follows therefore that $\bar{\chi}_0 \neq 1$, and hence necessarily $\bar{\chi}_0 = 0$. We deduce from (4.28) that $|\nabla u_0|^2 = x_2$ on $\partial \{u_0 > 0\}$. Computing the solution u_0 of the corresponding ordinary differential equation on ∂B_1 yields that

$$u_0(x) = \frac{\sqrt{2}}{3}\rho^{3/2}\cos(\frac{3}{2}(\min(\max(\theta, \frac{\pi}{6}), \frac{5\pi}{6}) - \frac{\pi}{2})), \quad \text{where } x = (\rho\cos\theta, \rho\sin\theta),$$

and that $M(0+) = \int_{B_1} x_2^+ \chi_{\{x:\pi/6<\theta<5\pi/6\}}\,dx$ in the case under consideration.

Consider now the case $u_0 = 0$. It follows from (4.27) that χ_0 is constant in $\{x_2 > 0\}$. Its value may be either 0 in which case $M(0+) = 0$, or 1 in which case $M(0+) = \int_{B_1} x_2^+ \, dx$.

Since the limit $M(0+)$ exists, the above proof shows that it can only take one of the three distinct values $\left\{0, \int_{B_1} x_2^+ \chi_{\{x:\pi/6<\theta<5\pi/6\}}\,dx, \int_{B_1} x_2^+ \,dx\right\}$. The above proof also yields, for each possible value of $M(0+)$, the existence of a *unique* blow-up limit, as claimed in the statement of the Theorem.

Under the assumption that the free boundary is locally an injective curve, we now derive its asymptotic behavior as it approaches a stagnation point.

Theorem 4 (Curve Case) *Let $n = 2$, let u be a weak solution of (4.12) satisfying Assumption 2, and let $x^0 \in \Omega$ be such that $x_2^0 = 0$. Suppose in addition that $\partial\{u > 0\}$ is in a neighborhood of x^0 a continuous injective curve $\sigma : (-t_0, t_0) \to \mathbf{R}^2$ such that $\sigma = (\sigma_1, \sigma_2)$ and $\sigma(0) = x^0$. Then the following hold:*

(i) *If $M(0+) = \int_{B_1} x_2^+ \chi_{\{x:\pi/6<\theta<5\pi/6\}}\,dx$, then (cf. Fig. 4.10) $\sigma_1(t) \neq x_1^0$ in $(-t_1, t_1) \setminus \{0\}$ and, depending on the parametrization, either*

$$\lim_{t\to 0+} \frac{\sigma_2(t)}{\sigma_1(t) - x_1^0} = \frac{1}{\sqrt{3}} \quad \text{and} \quad \lim_{t\to 0-} \frac{\sigma_2(t)}{\sigma_1(t) - x_1^0} = -\frac{1}{\sqrt{3}},$$

Fig. 4.10 Stokes corner

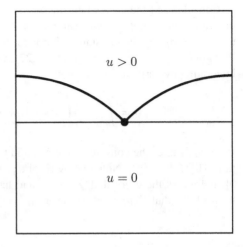

$u > 0$

$u = 0$

Fig. 4.11 Full density singularity

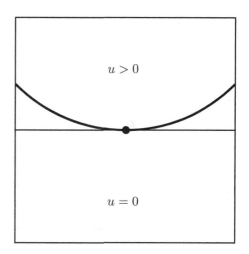

or

$$\lim_{t \to 0+} \frac{\sigma_2(t)}{\sigma_1(t) - x_1^0} = -\frac{1}{\sqrt{3}} \text{ and } \lim_{t \to 0-} \frac{\sigma_2(t)}{\sigma_1(t) - x_1^0} = \frac{1}{\sqrt{3}}.$$

(ii) *If* $M(0+) = \int_{B_1} x_2^+ \, dx$, *then (cf. Fig. 4.11)* $\sigma_1(t) \neq x_1^0$ *in* $(-t_1, t_1) \backslash \{0\}$, $\sigma_1 - x_1^0$ *changes sign at* $t = 0$ *and*

$$\lim_{t \to 0} \frac{\sigma_2(t)}{\sigma_1(t) - x_1^0} = 0.$$

(iii) *If* $M(0+) = 0$, *then (cf. Fig. 4.12)* $\sigma_1(t) \neq x_1^0$ *in* $(-t_1, t_1) \backslash \{0\}$, $\sigma_1 - x_1^0$ *does not change its sign at* $t = 0$, *and*

$$\lim_{t \to 0} \frac{\sigma_2(t)}{\sigma_1(t) - x_1^0} = 0.$$

Proof We may assume that $x_1^0 = 0$. Moreover, for each $y \in \mathbf{R}^2$ we define arg y as the complex argument of y, and we define the sets

$$\mathcal{L}_\pm := \{\theta_0 \in [0, \pi] : \text{ there is } t_m \to 0\pm \text{ such that } \arg \sigma(t_m) \to \theta_0 \text{ as } m \to \infty\}.$$

Step 1: *Both* $\mathcal{L}_+$ *and* $\mathcal{L}_-$ *are subsets of* $\{0, \pi/6, 5\pi/6, \pi\}$.

Indeed, suppose towards a contradiction that a sequence $0 \neq t_m \to 0, m \to \infty$ exists such that $\arg \sigma(t_m) \to \theta_0 \in (\mathcal{L}_+ \cup \mathcal{L}_-) \backslash \{0, \pi/6, 5\pi/6, \pi\}$, let $r_m := |\sigma(t_m)|$

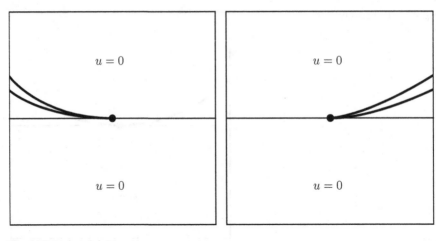

Fig. 4.12 Left and right cusp

and let

$$u_m(x) := \frac{u(r_m x)}{r_m^{3/2}}.$$

For each $\rho > 0$ such that $\tilde{B} := B_\rho(\cos\theta_0, \sin\theta_0)$ satisfies

$$\emptyset = \tilde{B} \cap \left(\{(x, 0) \; : \; x \in \mathbf{R}\} \cup \{(x, |x|/\sqrt{3}) \; : \; x \in \mathbf{R}\} \right),$$

we infer from the formula for the unique blow-up limit u_0 (see Theorem 3) that the signed measure

$$\Delta u_m(\tilde{B}) \to \Delta u_0(\tilde{B}) = 0 \text{ as } m \to \infty.$$

On the other hand,

$$\Delta u_m = -r_m^{1/2} f(r_m^{3/2} u_m) + \sqrt{x_2}\mathcal{H}^1 \lfloor \partial_{\{u_m>0\}},$$

which implies, since $\tilde{B} \cap \partial\{u_m > 0\}$ contains a curve of length at least $2\rho - o(1)$, that

$$0 \leftarrow \Delta u_m(\tilde{B}) \geq c(\theta_0, \rho) - C_1 r_m^{1/2} \text{ as } m \to \infty,$$

where $c(\theta_0, \rho) > 0$, a contradiction. Thus the property claimed in Step 1 holds.

Step 2: It follows that $\sigma_1(t) \neq 0$ for all sufficiently small $t \neq 0$. Now a continuity argument yields that both $\mathcal{L}_+$ and $\mathcal{L}_-$ are connected sets. Consequently

$$\ell_+ := \lim_{t \to 0+} \arg \sigma(t)$$

exists and is contained in the set $\{0, \pi/6, 5\pi/6, \pi\}$, and

$$\ell_- := \lim_{t \to 0-} \arg \sigma(t)$$

exists and is contained in the set $\{0, \pi/6, 5\pi/6, \pi\}$.

Step 3: In the case $M(0+) = \int_{B_1} x_2^+ \chi_{\{x:\pi/6<\theta<5\pi/6\}} dx$, we know now from the formula for u_0 that $\Delta u_0(B_{1/10}(\sqrt{3}/2, 1/2)) > 0$ and that $\Delta u_0(B_{1/10}(-\sqrt{3}/2, 1/2)) > 0$. It follows that the set $\{\ell_+, \ell_-\}$ contains both $\pi/6$ and $5\pi/6$. But then the sets $\{\ell_+, \ell_-\}$ and $\{\pi/6, 5\pi/6\}$ must be equal, and the fact that $u = 0$ on $x_2 = 0$ implies case (i) of the Theorem.

Step 4: In the case $M(0+) \in \{0, \int_{B_1} x_2^+ dx\}$, we have that $\Delta u_0(B_{1/10}(\sqrt{3}/2, 1/2)) = 0$, which implies that $\ell_+, \ell_- \notin \{\pi/6, 5\pi/6\}$. Thus $\ell_+, \ell_- \in \{0, \pi\}$. Using the fact that $u = 0$ on $x_2 = 0$, we obtain in the case $\ell_+ \neq \ell_-$ that $M(0+) = \int_{B_1} x_2^+ dx$ and in the case $\ell_+ = \ell_-$ that $M(0+) = 0$. Together, the last two properties prove case (ii) and case (iii) of the Theorem.

Remark 3 In [37] we used a strong version of the Rayleigh-Taylor condition (which is always valid in the case of zero vorticity) in order to prove that the cusps of case (iii) are not possible. Unfortunately we do not have the Rayleigh-Taylor condition $|\nabla u|^2 + 2F(u) - x_n^+ \leq 0$ in the case with nonzero vorticity, and the method of [37] breaks down here. Still we *conjecture* that the cusps in case (iii) are *not* possible when assuming the Rayleigh-Taylor condition.

4.8 Definition of Degenerate Points

Definition 3 (Stagnation Points) Let u be a variational solution of (4.12). We call $S^u := \{x \in \Omega : x_n = 0 \text{ and } x \in \partial\{u > 0\}\}$ the set of *stagnation points*.

Throughout the rest of this section we assume that $n = 2$.

Definition 4 (Non-degeneracy) Let u be a variational solution of (4.12). We say that a point $x^0 \in \Omega \cap \partial\{u > 0\} \cap \{x_2 = 0\}$ is *degenerate* if

$$\frac{u(x^0 + rx)}{r^{3/2}} \to 0 \quad \text{as } r \to 0+,$$

strongly in $W_{loc}^{1,2}(\mathbf{R}^2)$. Otherwise we call $x^0 \in \Omega \cap \partial\{u > 0\} \cap \{x_2 = 0\}$ *non-degenerate*.

Remark 4 Note that Theorem 3 gives alternative characterizations of degenera-cy/non-degeneracy in terms of the blow-up limit or the density.

Remark 5 It is possible to show that under certain assumptions, degenerate points are isolated (see [38, Proposition 5.4]).

Definition 5 Let u be a variational solution of (4.12). We define

$$\Sigma^u := \{x^0 \in S^u : M_{x^0,u}(0+) = \int_{B_1} x_n^+ \, dx\}.$$

Remark 6 The set Σ^u is closed, as a consequence of the upper semicontinuity Lemma 1(v) as well as the characterization of the set of values of $M_{x^0,u}(0+)$ in Theorem 3.

Remark 7 In the case of two dimensions and $\{u > 0\}$ being a supergraph or a Lipschitz set (each of the latter assumptions excluding the case $M_{x^0,u}(0+) = 0$), we infer from Theorem 3 that the set $S^u \setminus \Sigma^u$ equals the set of non-degenerate stagnation points and is according to the previous remark a finite or countable set.

Remark 8

(i) In the case $-f \equiv c > 0$, the function $u(x) = \frac{c}{2}x_n^2$ is a weak solution of (4.12). In this example, $\Sigma^u = \{x_n = 0\}$. Similarly, one may prove that for any f such that $f(0) < 0$, there exists an explicit solution $u(x) = u(x_n)$ such that $\Sigma^u = \{x_n = 0\}$. Thus degenerate points may exist in the case $f(0) < 0$.

(ii) [36, Proposition 5.12] shows that $\Sigma^u = \emptyset$ in the case when $n = 2$ and $f \geq 0$ in a right neighborhood of 0 (in particular this is satisfied when $f(0) > 0$).

4.9   Frequency Formula

Frequency Formulae at points of highest density had been used in [42] and [37]. In the presence of vorticity it is more complicated:

Theorem 5 (Frequency Formula) *Let u be a variational solution of (4.12) sat-isfying Assumption 2, let x^0 be a stagnation point, and let $\delta := \text{dist}(x^0, \partial\Omega)/2$. Let*

$$D_{x^0,u}(r) = D(r) = \frac{r \int_{B_r(x^0)} (|\nabla u|^2 - uf(u)) \, dx}{\int_{\partial B_r(x^0)} u^2 \, d\mathcal{H}^{n-1}}$$

and

$$V_{x^0,u}(r) = V(r) = \frac{r \int_{B_r(x^0)} x_n^+ (1 - \chi_{\{u>0\}}) \, dx}{\int_{\partial B_r(x^0)} u^2 \, d\mathcal{H}^{n-1}}.$$

Then the "frequency"

$$H_{x^0,u}(r) = H(r) = D(r) - V(r)$$

$$= \frac{r \int_{B_r(x^0)} \left(|\nabla u|^2 - u f(u) + x_n^+ (\chi_{\{u>0\}} - 1) \right) dx}{\int_{\partial B_r(x^0)} u^2 \, d\mathcal{H}^{n-1}}$$

satisfies for a.e. $r \in (0, \delta)$ the identities

$H'(r)$

$$= \frac{2}{r} \int_{\partial B_r(x^0)} \left[\frac{r(\nabla u \cdot v)}{\left(\int_{\partial B_r(x^0)} u^2 \, d\mathcal{H}^{n-1} \right)^{1/2}} - D(r) \frac{u}{\left(\int_{\partial B_r(x^0)} u^2 \, d\mathcal{H}^{n-1} \right)^{1/2}} \right]^2 d\mathcal{H}^{n-1}$$

$$+ \frac{2}{r} V^2(r) + \frac{2}{r} V(r) \left(H(r) - \frac{3}{2} \right) + \frac{K(r)}{\int_{\partial B_r(x^0)} u^2 \, d\mathcal{H}^{n-1}} \tag{4.29}$$

and

$H'(r)$

$$= \frac{2}{r} \int_{\partial B_r(x^0)} \left[\frac{r(\nabla u \cdot v)}{\left(\int_{\partial B_r(x^0)} u^2 \, d\mathcal{H}^{n-1} \right)^{1/2}} - H(r) \frac{u}{\left(\int_{\partial B_r(x^0)} u^2 \, d\mathcal{H}^{n-1} \right)^{1/2}} \right]^2 d\mathcal{H}^{n-1}$$

$$+ \frac{2}{r} V(r) \left(H(r) - \frac{3}{2} \right) + \frac{K(r)}{\int_{\partial B_r(x^0)} u^2 \, d\mathcal{H}^{n-1}}; \tag{4.30}$$

here

$$K(r) = r \int_{\partial B_r(x^0)} (2F(u) - u f(u)) \, d\mathcal{H}^{n-1}$$

$$+ \int_{B_r(x^0)} ((n-2) u f(u) - 2n F(u)) \, dx$$

is the function defined in Theorem 1.

Remark 9 The root of this formula is the classical frequency formula of F. Almgren for Q-valued harmonic functions [1]. Almgren's formula has subsequently been extended to various perturbations. Note however that while our formula may look like a perturbation of the "linear" formula for Q-valued harmonic functions, it is in fact a truly nonlinear formula.

Proof Note that, for all $r \in (0, \delta)$,

$$H(r) = \frac{I(r) - \int_{B_1} x_n^+ \, dx}{J(r)}.$$

Hence, for a.e. $r \in (0, \delta)$,

$$H'(r) = \frac{I'(r)}{J(r)} - \frac{(I(r) - \int_{B_1} x_n^+ \, dx) \, J'(r)}{J(r) \quad J(r)},$$

Using the identities (4.17) and (4.18), we therefore obtain that, for a.e. $r \in (0, \delta)$,

$$H'(r) = \frac{\left(2r \int_{\partial B_r(x^0)} (\nabla u \cdot v)^2 \, d\mathcal{H}^{n-1} - 3 \int_{\partial B_r(x^0)} u \nabla u \cdot v \, d\mathcal{H}^{n-1}\right) + K(r)}{\int_{\partial B_r(x^0)} u^2 \, d\mathcal{H}^{n-1}}$$

$$- (D(r) - V(r)) \frac{1}{r} \frac{\left(2r \int_{\partial B_r(x^0)} u \nabla u \cdot v \, d\mathcal{H}^{n-1} - 3 \int_{\partial B_r(x^0)} u^2 \, d\mathcal{H}^{n-1}\right)}{\int_{\partial B_r(x^0)} u^2 \, d\mathcal{H}^{n-1}}$$

$$= \frac{2}{r} \left(\frac{r^2 \int_{\partial B_r(x^0)} (\nabla u \cdot v)^2 \, d\mathcal{H}^{n-1}}{\int_{\partial B_r(x^0)} u^2 \, d\mathcal{H}^{n-1}} - \frac{3}{2} D(r)\right)$$

$$- \frac{2}{r} (D(r) - V(r)) \left(D(r) - \frac{3}{2}\right) + \frac{K(r)}{\int_{\partial B_r(x^0)} u^2 \, d\mathcal{H}^{n-1}}, \tag{4.31}$$

where we have also used the fact, which follows from (4.21), that

$$D(r) = \frac{r \int_{\partial B_r(x^0)} u \nabla u \cdot v \, d\mathcal{H}^{n-1}}{\int_{\partial B_r(x^0)} u^2 \, d\mathcal{H}^{n-1}}. \tag{4.32}$$

Identity (4.29) now follows by merely rearranging (4.31), making use again of (4.32) and the fact that $D(r) = V(r) + H(r)$.

Since (4.29) holds, it follows by inspection that (4.30) holds if and only if

$$\int_{\partial B_r(x^0)} [r(\nabla u \cdot v) - D(r)u]^2 \, d\mathcal{H}^{n-1} + V^2(r) \int_{\partial B_r(x^0)} u^2 \, d\mathcal{H}^{n-1}$$

$$= \int_{\partial B_r(x^0)} [r(\nabla u \cdot v) - H(r)u]^2 \, d\mathcal{H}^{n-1}. \tag{4.33}$$

However, (4.33) is easily verified as a consequence of (4.32) and the fact that $D(r) = H(r) + V(r)$. In conclusion, identity (4.30) also holds.

The following lemma is motivated by [19, (4.11)].

Lemma 2 *Let u be a variational solution of (4.12). Then*

$$r \int_{\partial B_r(x^0)} u^2 \, d\mathcal{H}^{n-1} = \int_{B_r(x^0)} \left(nu^2 + (|\nabla u|^2 - uf(u))(r^2 - |x|^2) \right) dx.$$
(4.34)

Proof As

$$\int_{B_r(x^0)} 2nu^2 \, dx = -\int_{B_r(x^0)} u^2 \Delta(r^2 - |x|^2) \, dx,$$

a proof can be obtained integrating by parts twice.

From now on we make the following assumption concerning the growth of f:

Assumption 6 *There exists a constant $C < +\infty$ such that*

$$|f(z)| \le Cz \quad \text{for all } z \in (0, z_0).$$
(4.35)

Note that when f is a C^1 function, the above is a consequence of $f(0) = 0$. Assumption 6 also implies that

$$|F(z)| \le Cz^2/2 \quad \text{for all } z \in (0, z_0).$$

As a corollary of Lemma 2 we obtain thus:

Corollary 1 *Let u be a variational solution of (4.12) such that Assumptions 2 and 6 hold. Then there exists $r_0 > 0$ such that*

$$r \int_{\partial B_r(x^0)} u^2 \, d\mathcal{H}^{n-1} \ge \int_{B_r(x^0)} u^2 \quad \text{for all } r \in (0, r_0)$$

and

$$|K(r)| \le C_0 r \int_{\partial B_r(x^0)} u^2 \quad \text{for all } r \in (0, r_0).$$
(4.36)

Theorem 7 *Let u be a variational solution of (4.12) such that Assumptions 2 and 6 hold, let $x^0 \in \Sigma^u$, and let $\delta := \text{dist}(x^0, \partial\Omega)/2$. Then the following hold, for some $r_0 \in (0, \delta)$ sufficiently small:*

(i) *There exists a positive constant C_1 such that*

$$H(r) - \frac{3}{2} \ge -C_1 r^2 \quad \text{for all } r \in (0, r_0).$$

(ii) *There exists a positive constant β such that*

$$r \mapsto e^{\beta r^2} J(r) \quad \text{is nondecreasing on } (0, r_0).$$

(iii) $r \mapsto \frac{1}{r} V^2(r) \in L^1(0, r_0)$.
(iv) *The function H has a right limit H(0+), where H(0+) ≥ 3/2.*
(v) *The function*

$$H'(r) - \frac{2}{r} \int_{\partial B_r(x^0)} \left[\frac{r(\nabla u \cdot v)}{\left(\int_{\partial B_r(x^0)} u^2 \, d\mathcal{H}^{n-1} \right)^{1/2}} - H(r) \frac{u}{\left(\int_{\partial B_r(x^0)} u^2 \, d\mathcal{H}^{n-1} \right)^{1/2}} \right]^2 d\mathcal{H}^{n-1}$$

is bounded from below by a function in $L^1(0, r_0)$.

Proof Since assumption (4.35) holds, we deduce from (4.19) using (4.36) that, for all r sufficiently small,

$$I(r) - \frac{3}{2} J(r) - \int_{B_1} x_n^+ \, dx \geq -C_0 \int_0^r t^{-n-1} \int_{\partial B_t(x^0)} u^2 \, d\mathcal{H}^{n-1} dt. \tag{4.37}$$

This implies that, for all $r \in (0, r_0)$,

$$r^{-n-1} \int_{B_r} (|\nabla u|^2 - u f(u)) \, dx - \frac{3}{2} r^{-n-2} \int_{\partial B_r(x^0)} u^2 \, d\mathcal{H}^{n-1}$$

$$\geq -C_0 \int_0^r t^{-n-1} \int_{\partial B_t(x^0)} u^2 \, d\mathcal{H}^{n-1} dt. \tag{4.38}$$

Let $Y : (0, r_0) \to \mathbf{R}$ be given by

$$Y(r) = \int_0^r t^{-n-1} \int_{\partial B_t(x^0)} u^2 \, d\mathcal{H}^{n-1} dt.$$

We deduce from (4.18) and (4.38) that

$$\frac{d}{dr} \left(\frac{Y'(r)}{r} \right) \geq -\alpha \frac{Y(r)}{r}, \tag{4.39}$$

for some positive constant $\alpha < +\infty$. Observe now that, as a consequence of the *Bessel type differential inequality* (4.39),

$$\frac{d^2}{dr^2} \left(\frac{Y(r)}{r^{1/2}} \right) \geq \frac{\frac{3}{4} - \alpha r^2}{r^{5/2}} Y(r) \geq 0 \quad \text{for all } r \in (0, r_0), \tag{4.40}$$

for some r_0 sufficiently small. Thus $r \mapsto Y(r)/r^{1/2}$ is a convex function on $(0, r_0)$, and since

$$\lim_{r \to 0+} \frac{Y(r)}{r^{1/2}} = 0,$$

it follows that

$$\frac{Y(r)}{r^{1/2}} - 0 \le (r - 0)\frac{d}{dr}\left(\frac{Y(r)}{r^{1/2}}\right) \quad \text{for all } r \in (0, r_0),$$

and therefore

$$\frac{3}{2}\frac{Y(r)}{r} \le Y'(r) \quad \text{for all } r \in (0, r_0).$$

This implies, together with (4.37), that

$$r^{-n-1}\int_{B_r}(|\nabla u|^2 - uf(u)) - x_n^+(1 - \chi_{\{u>0\}})\, dx - \frac{3}{2}r^{-n-2}\int_{\partial B_r} u^2\, d\mathcal{H}^{n-1}$$

$$\ge -\frac{2}{3}C_0 r^{-n}\int_{\partial B_r(x^0)} u^2\, d\mathcal{H}^{n-1}, \tag{4.41}$$

which is equivalent to (i).

Taking also into account (4.18), (4.41) also implies that, for a.e. r sufficiently small,

$$J'(r) \ge -2\beta r J(r),$$

for some constant $\beta > 0$, which is equivalent to (ii).

Now, using (4.36) and part (i) in (4.29), we obtain that, for a.e. $r \in (0, r_0)$,

$$H'(r) \ge \frac{2}{r}V^2(r) - 2C_1 r V(r) - C_0 r. \tag{4.42}$$

As

$$2C_1 r V(r) \le \frac{1}{r}V^2(r) + C_1^2 r^3, \tag{4.43}$$

we obtain from (4.42) that, for a.e. $r \in (0, r_0)$,

$$H'(r) \ge \frac{1}{r}V^2(r) - C_1^2 r^3 - C_0 r. \tag{4.44}$$

Since, by part (ii), $r \mapsto H(r)$ is bounded below as $r \to 0$, we obtain (iii). We also deduce from (4.44) and part (i) that $H(r)$ has a limit as $r \to 0+$, and that $H(0+) \geq 3/2$, thus proving (iv).

We now consider (4.30), and deduce from part (i) using (4.43) that, for a.e. $r \in (0, r_0)$,

$$
H'(r) - \frac{2}{r} \int_{\partial B_r(x^0)} \left[\frac{r(\nabla u \cdot \nu)}{\left(\int_{\partial B_r(x^0)} u^2 \, d\mathcal{H}^{n-1} \right)^{1/2}} \right.
$$

$$
\left. - H(r) \frac{u}{\left(\int_{\partial B_r(x^0)} u^2 \, d\mathcal{H}^{n-1} \right)^{1/2}} \right]^2 d\mathcal{H}^{n-1}
$$

$$
\geq -2C_0 r \, V(r) - C_2 r
$$

$$
\geq -\frac{1}{r} V^2(r) - C_1^2 r^3 - C_0 r, \tag{4.46}
$$

which, together with part (iii), proves (v).

4.10 Blow-Up Limits

The Frequency Formula allows passing to blow-up limits.

Proposition 1 *Let u be a variational solution of (4.12), and let $x^0 \in \Sigma^u$. Then:*

(i) *There exist $\lim_{r \to 0+} V(r) = 0$ and $\lim_{r \to 0+} D(r) = H_{x^0,u}(0+)$.*
(ii) *For any sequence $r_m \to 0+$ as $m \to \infty$, the sequence*

$$
v_m(x) := \frac{u(x^0 + r_m x)}{\sqrt{r_m^{1-n} \int_{\partial B_{r_m}(x^0)} u^2 \, d\mathcal{H}^{n-1}}} \tag{4.47}
$$

is bounded in $W^{1,2}(B_1)$.
(iii) *For any sequence $r_m \to 0+$ as $m \to \infty$ such that the sequence v_m in (4.47) converges weakly in $W^{1,2}(B_1)$ to a blow-up limit v_0, the function v_0 is homogeneous of degree $H_{x^0,u}(0+)$ in B_1, and satisfies*

$$
v_0 \geq 0 \text{ in } B_1, \quad v_0 \equiv 0 \text{ in } B_1 \cap \{x_n \leq 0\} \text{ and } \int_{\partial B_1} v_0^2 \, d\mathcal{H}^{n-1} = 1.
$$

Proof We first prove that, for any sequence $r_m \to 0+$, the sequence v_m defined in (4.47) satisfies, for every $0 < \varrho < \sigma < 1$,

$$\int_{B_\sigma \setminus B_\varrho} |x|^{-n-3} \left[\nabla v_m(x) \cdot x - H_{x^0,u}(0+)v_m(x) \right]^2 dx \to 0 \quad \text{as } m \to \infty.$$

(4.48)

Indeed, for any such ϱ and σ, it follows by scaling from (4.46) that, for every m such that $r_m < \delta$,

$$\int_\varrho^\sigma \frac{2}{r} \int_{\partial B_r} \left[\frac{r(\nabla v_m \cdot \nu)}{\left(\int_{\partial B_r} v_m^2 \, d\mathcal{H}^{n-1} \right)^{1/2}} - H(r_m r) \frac{v_m}{\left(\int_{\partial B_r} v_m^2 \, d\mathcal{H}^{n-1} \right)^{1/2}} \right]^2 d\mathcal{H}^{n-1} \, dr$$

$$\leq H(r_m \sigma) - H(r_m \varrho) + \int_{r_m \varrho}^{r_m \sigma} \frac{1}{r} V^2(r) + C_1^2 r^3 + C_0 r \, dr \to 0 \quad \text{as } m \to \infty,$$

as a consequence of Theorem 7 (iv)–(v). The above implies that

$$\int_\varrho^\sigma \frac{2}{r} \int_{\partial B_r} \left[\frac{r(\nabla v_m \cdot \nu)}{\left(\int_{\partial B_r} v_m^2 \, d\mathcal{H}^{n-1} \right)^{1/2}} - H(0+) \frac{v_m}{\left(\int_{\partial B_r} v_m^2 \, d\mathcal{H}^{n-1} \right)^{1/2}} \right]^2 d\mathcal{H}^{n-1} \, dr$$

$$\to 0 \quad \text{as } m \to \infty.$$

(4.49)

Now note that, for every $r \in (\varrho, \sigma) \subset (0, 1)$ and all m as before, it follows by using Theorem 7(ii), that

$$\int_{\partial B_r} v_m^2 \, d\mathcal{H}^{n-1} = \frac{\int_{\partial B_{r_m r}(x_0)} u^2 \, d\mathcal{H}^{n-1}}{\int_{\partial B_{r_m}(x_0)} u^2 \, d\mathcal{H}^{n-1}} \leq e^{\beta r_m^2 (1-r^2)} r^{n+2} \to r^{n+2}, m \to \infty.$$

Therefore (4.48) follows from (4.49), which proves our claim.

Let us also note that, as a consequence of Corollary 1, for each r sufficiently small

$$\left| D(r) - \frac{r \int_{B_r(x^0)} |\nabla u|^2 \, dx}{\int_{\partial B_r(x^0)} u^2 \, d\mathcal{H}^{n-1}} \right| \leq Cr^2.$$

(4.50)

This implies that, for any sequence $r_m \to 0+$, the sequence v_m defined in (4.47) satisfies

$$\left| D(r_m) - \int_{B_1} |\nabla v_m|^2 \, dx \right| \leq Cr_m^2.$$

(4.51)

We can now prove all parts of the Proposition.

(i) Suppose towards a contradiction that (i) is not true. Let $s_m \to 0$ be such that the sequence $V(s_m)$ is bounded away from 0. From the integrability of $r \mapsto \frac{2}{r} V^2(r)$ we obtain that

$$\min_{r \in [s_m, 2s_m]} V(r) \to 0 \quad \text{as } m \to \infty.$$

Let $t_m \in [s_m, 2s_m]$ be such that $V(t_m) \to 0$ as $m \to \infty$. For the choice $r_m := t_m$ for every m, the sequence v_m given by (4.47) satisfies (4.48). The fact that $V(r_m) \to 0$ implies that $D(r_m)$ is bounded, and hence, using (4.51), that v_m is bounded in $W^{1,2}(B_1)$. Let v_0 be any weak limit of v_m along a subsequence. Note that by the compact embedding $W^{1,2}(B_1) \hookrightarrow L^2(\partial B_1)$, v_0 has norm 1 on $L^2(\partial B_1)$, since this is true for v_m for all m. It follows from (4.48) that v_0 is homogeneous of degree $H_{x^0, u}(0+)$. Note that, by using Theorem 7(ii),

$$
\begin{aligned}
V(s_m) &= \frac{s_m^{-n-1} \int_{B_{s_m}(x^0)} x_n^+ (1 - \chi_{\{u>0\}}) \, dx}{s_m^{-n-2} \int_{\partial B_{s_m}(x^0)} u^2 \, d\mathcal{H}^{n-1}} \\
&\leq \frac{s_m^{-n-1} \int_{B_{r_m}(x^0)} x_n^+ (1 - \chi_{\{u>0\}}) \, dx}{e^{\beta[(r_m^2/4) - s_m^2]} (r_m/2)^{-n-2} \int_{\partial B_{r_m/2}(x^0)} u^2 \, d\mathcal{H}^{n-1}} \\
&\leq \frac{e^{3\beta r_m^2/4} \int_{\partial B_{r_m}(x^0)} u^2 \, d\mathcal{H}^{n-1}}{2 \int_{\partial B_{r_m/2}(x^0)} u^2 \, d\mathcal{H}^{n-1}} V(r_m) \\
&= \frac{e^{3\beta r_m^2/4}}{2 \int_{\partial B_{1/2}} v_m^2 \, d\mathcal{H}^{n-1}} V(r_m).
\end{aligned}
\tag{4.52}
$$

Since, at least along a subsequence,

$$\int_{\partial B_{1/2}} v_m^2 \, d\mathcal{H}^{n-1} \to \int_{\partial B_{1/2}} v_0^2 \, d\mathcal{H}^{n-1} > 0,$$

Equation (4.52) leads to a contradiction. It follows that indeed $V(r) \to 0$ as $r \to 0+$. This implies that $D(r) \to H_{x^0, u}(0+)$.

(ii) Let r_m be an arbitrary sequence with $r_m \to 0+$. In view of (4.51), the boundedness of the sequence v_m in $W^{1,2}(B_1)$ is equivalent to the boundedness of $D(r_m)$, which is true by (i).

(iii) Let $r_m \to 0+$ be an arbitrary sequence such that v_m converges weakly to v_0. The homogeneity degree $H_{x^0, u}(0+)$ of v_0 follows directly from (4.48). The fact that $\int_{\partial B_1} v_0^2 \, d\mathcal{H}^{n-1} = 1$ is a consequence of $\int_{\partial B_1} v_m^2 \, d\mathcal{H}^{n-1} = 1$ for all m, and the remaining claims of the Proposition are obvious.

4.11 Concentration Compactness in Two Dimensions

In the two-dimensional case we prove concentration compactness which allows us to preserve variational solutions in the blow-up limit at degenerate points and excludes concentration. In order to do so we combine the concentration compactness result of Evans and Müller [14] with information gained by our Frequency Formula. In addition, we obtain strong convergence of our blow-up sequence which is necessary in order to prove our main theorems. The question whether the following Theorem holds in any dimension seems to be a hard one.

Theorem 8 *Let* $n = 2$, *let the nonlinearity satisfy Assumption 6 and let u be a variational solution of* (4.12), *and let* $x^0 \in \Sigma^u$. *Let* $r_m \to 0+$ *be such that the sequence* v_m *given by* (4.47) *converges weakly to* v_0 *in* $W^{1,2}(B_1)$. *Then* v_m *converges to* v_0 *strongly in* $W_{loc}^{1,2}(B_1 \setminus \{0\})$, v_0 *is continuous on* B_1 *and* Δv_0 *is a nonnegative Radon measure satisfying* $v_0 \Delta v_0 = 0$ *in the sense of Radon measures in* B_1.

Proof The proof is similar to that in [37, Theorem 9.1], but there are some subtle changes so that we will supply the whole proof for the sake of completeness.

Note first that the homogeneity of v_0 given by Proposition 1, together with the fact that v_0 belongs to $W^{1,2}(B_1)$, imply that v_0 is continuous. As

$$\Delta v_m(x) = \frac{r_m^2 \Delta u(x^0 + r_m x)}{\sqrt{r_m^{-1} \int_{\partial B_{r_m}(x^0)} u^2 \, d\mathcal{H}^{n-1}}} = \frac{-r_m^2 f(u(x^0 + r_m x))}{\sqrt{r_m^{-1} \int_{\partial B_{r_m}(x^0)} u^2 \, d\mathcal{H}^{n-1}}} \qquad (4.53)$$

$$\geq -C_1 \frac{-r_m^2 u(x^0 + r_m x)}{\sqrt{r_m^{-1} \int_{\partial B_{r_m}(x^0)} u^2 \, d\mathcal{H}^{n-1}}} = -C_1 r_m^2 v_m(x) \text{ for } v_m(x) > 0,$$

we obtain from the sign of the singular part of Δv_m with respect to the Lebesgue measure that $\Delta v_m \geq -C_1 r_m^2 v_m$ in B_1 in the sense of measures. From [20, Theorem 8.17] we infer therefore that

$$\sup_{B_\sigma} v_m \leq C_2(\sigma) \int_{B_1} v_m \, dx$$

for each $\sigma \in (0, 1)$. Consequently

$$\Delta v_m \geq -C_3(\sigma) r_m^2 \text{ in } B_\sigma \qquad (4.54)$$

in the sense of measures. It follows that for each nonnegative $\eta \in C_0^\infty(B_1)$ such that $\eta = 1$ in $B_{(\sigma+1)/2}$

$$\int_{B_{(\sigma+1)/2}} d\Delta v_m = \int_{B_{(\sigma+1)/2}} \eta \, d\Delta v_m \leq \int_{B_1} \eta \, d\Delta v_m + C_1 r_m^2 \int_{B_1 \setminus B_{(\sigma+1)/2}} v_m$$

$$= \int_{B_1} v_m \Delta \eta + C_1 r_m^2 \int_{B_1 \setminus B_{(\sigma+1)/2}} v_m \leq C_4 \text{ for all } m \in \mathbf{N}.$$

(4.55)

From (4.53) and the fact that v_m is bounded in $L^1(B_1)$, we obtain also that Δv_0 is a nonnegative Radon measure on B_1. The continuity of v_0 implies therefore that $v_0 \Delta v_0$ is well defined as a nonnegative Radon measure on B_1.

In order to apply the concentrated compactness result [14], we modify each v_m to

$$\tilde{v}_m := (v_m + C_3(\sigma) r_m^2 |x|^2) * \phi_m \in C^\infty(B_1),$$

where ϕ_m is a standard mollifier such that

$$\Delta \tilde{v}_m \geq 0, \int_{B_\sigma} d\Delta \tilde{v}_m \leq C_2 < +\infty \quad \text{for all } m,$$

and

$$\|v_m - \tilde{v}_m\|_{W^{1,2}(B_\sigma)} \to 0 \quad \text{as } m \to \infty.$$

From [15, Chapter 4, Theorem 3] we know that $\nabla \tilde{v}_m$ converges a.e. to the weak limit ∇v_0, and the only possible problem is concentration of $|\nabla \tilde{v}_m|^2$. By [14, Theorem 1.1] and [14, Theorem 3.1] we obtain that

$$\partial_1 \tilde{v}_m \partial_2 \tilde{v}_m \to \partial_1 v_0 \partial_2 v_0$$

and

$$(\partial_1 \tilde{v}_m)^2 - (\partial_2 \tilde{v}_m)^2 \to (\partial_1 v_0)^2 - (\partial_2 v_0)^2$$

in the sense of distributions on B_σ as $m \to \infty$. It follows that

$$\partial_1 v_m \partial_2 v_m \to \partial_1 v_0 \partial_2 v_0$$

(4.56)

and

$$(\partial_1 v_m)^2 - (\partial_2 v_m)^2 \to (\partial_1 v_0)^2 - (\partial_2 v_0)^2$$

in the sense of distributions on B_σ as $m \to \infty$. Let us remark that this alone would allow us to pass to the limit in the domain variation formula for v_m in the set $\{x_2 > 0\}$.

Observe now that (4.48) shows that for each $0 < \varrho < \sigma$

$$\nabla v_m(x) \cdot x - H_{x^0,u}(0+)v_m(x) \to 0$$

strongly in $L^2(B_\sigma \setminus B_\varrho)$ as $m \to \infty$. It follows that

$$\partial_1 v_m x_1 + \partial_2 v_m x_2 \to \partial_1 v_0 x_1 + \partial_2 v_0 x_2$$

strongly in $L^2(B_\sigma \setminus B_\varrho)$ as $m \to \infty$. But then

$$\int_{B_\sigma \setminus B_\varrho} (\partial_1 v_m \partial_1 v_m x_1 + \partial_1 v_m \partial_2 v_m x_2)\eta \, dx$$

$$\to \int_{B_\sigma \setminus B_\varrho} (\partial_1 v_0 \partial_1 v_0 x_1 + \partial_1 v_0 \partial_2 v_0 x_2)\eta \, dx$$

for each $\eta \in C_0^0(B_\sigma \setminus \overline{B}_\varrho)$ as $m \to \infty$. Using (4.56), we obtain that

$$\int_{B_\sigma \setminus B_\varrho} (\partial_1 v_m)^2 x_1 \eta \, dx \to \int_{B_\sigma \setminus B_\varrho} (\partial_1 v_0)^2 x_1 \eta \, dx$$

for each $0 \le \eta \in C_0^0((B_\sigma \setminus \overline{B}_\varrho) \cap \{x_1 > 0\})$ and for each $0 \ge \eta \in C_0^0((B_\sigma \setminus \overline{B}_\varrho) \cap \{x_1 < 0\})$ as $m \to \infty$. Repeating the above procedure three times for rotated sequences of solutions (by 45 degrees) yields that ∇v_m converges strongly in $L_{\text{loc}}^2(B_\sigma \setminus \overline{B}_\varrho)$. Since σ and ϱ with $0 < \varrho < \sigma < 1$ were arbitrary, it follows that ∇v_m converges to ∇v_0 strongly in $L_{\text{loc}}^2(B_1 \setminus \{0\})$.

As a consequence of the strong convergence and Assumption 6, we obtain now, using the fact that the singular part of Δv_m lives on a subset of $\{v_m = 0\}$, that

$$\left| \int_{B_1} \nabla(\eta v_0) \cdot \nabla v_0 \, dx \right| \leftarrow \left| \int_{B_1} \nabla(\eta v_m) \cdot \nabla v_m \, dx \right|$$

$$\le C_1 r_m^2 \int_{B_1} \eta v_m^2 \, dx \to 0, m \to \infty \quad \text{for all } \eta \in C_0^1(B_1 \setminus \{0\}).$$

Combined with the fact that $v_0 = 0$ in $B_1 \cap \{x_2 \le 0\}$ and the fact that the singular part of Δv_0 lives on a subset of $\{v_0 = 0\} \cup \{x_2 = 0\}$, this proves that $v_0 \Delta v_0 = 0$ in the sense of Radon measures on B_1.

4.12 Degenerate Points in Two Dimensions

Theorem 9 *Let $n = 2$, let the nonlinearity satisfy Assumption 6 and let u be a variational solution of (4.12). Then at each point x^0 of the set Σ^u there exists an integer $N(x^0) \geq 2$ such that*

$$H_{x^0,u}(0+) = N(x^0)$$

and

$$\frac{u(x^0 + rx)}{\sqrt{r^{-1} \int_{\partial B_r(x^0)} u^2 \, d\mathcal{H}^1}} \to \frac{\rho^{N(x^0)} |\sin(N(x^0) \min(\max(\theta, 0), \pi))|}{\sqrt{\int_0^\pi \sin^2(N(x^0)\theta)d\theta}} \quad \text{as } r \to 0+,$$

strongly in $W^{1,2}_{loc}(B_1 \setminus \{0\})$ and weakly in $W^{1,2}(B_1)$, where $x = (\rho \cos \theta, \rho \sin \theta)$.

Proof Let $r_m \to 0+$ be an arbitrary sequence such that the sequence v_m given by (4.47) converges weakly in $W^{1,2}(B_1)$ to a limit v_0. By Proposition 1(iii) and Theorem 8, $v_0 \not\equiv 0$, v_0 is homogeneous of degree $H_{x^0,u}(0+) \geq 3/2$, v_0 is continuous, $v_0 \geq 0$ and $v_0 \equiv 0$ in $\{x_2 \leq 0\}$, $v_0 \Delta v_0 = 0$ in B_1 as a Radon measure, and the convergence of v_m to v_0 is strong in $W^{1,2}_{loc}(B_1 \setminus \{0\})$. Moreover, the strong convergence of v_m and the fact proved in Proposition 1(i) that $V(r_m) \to 0$ as $m \to \infty$ imply that

$$0 = \int_{B_1} \left(|\nabla v_0|^2 \text{div} \, \phi - 2\nabla v_0 D\phi \nabla v_0 \right) dx$$

for every $\phi \in C_0^1(B_1 \cap \{x_2 > 0\}; \mathbf{R}^2)$. It follows that at each polar coordinate point $(1, \theta) \in \partial B_1 \cap \partial \{v_0 > 0\}$,

$$\lim_{\tau \to \theta+} \partial_\theta v_0(1, \tau) = - \lim_{\tau \to \theta-} \partial_\theta v_0(1, \tau).$$

Computing the solution of the ODE on ∂B_1, using the homogeneity of degree $H_{x^0,u}(0+)$ of v_0 and the fact that $\int_{\partial B_1} v_0^2 \, d\mathcal{H}^1 = 1$, yields that $H_{x^0,u}(0+)$ must be an *integer* $N(x^0) \geq 2$ and that

$$v_0(\rho, \theta) = \frac{\rho^{N(x^0)} |\sin(N(x^0) \min(\max(\theta, 0), \pi))|}{\sqrt{\int_0^\pi \sin^2(N(x^0)\theta)d\theta}}. \tag{4.57}$$

The desired conclusion follows from Proposition 1(ii).

Remark 10 One may show ([38, Theorem 9.2]) that in dimension 2 the set Σ^u is locally in Ω a finite set.

4.13 Main Reuslt

Theorem 10 *Let* $n = 2$, *let* u *be a weak solution of* (4.12) *satisfying Assumption* 2, *let the free boundary* $\partial\{u > 0\}$ *be a continuous injective curve* $\sigma = (\sigma_1, \sigma_2)$ *such that* $\sigma(0) = x^0 = (x_1^0, 0)$, *and assume that the nonlinearity* f *satisfies either Assumption* 6, *or* $f \geq 0$ *in a right neighborhood of* 0.

(i) *If* $M_{x^0,u}(0+) = \int_{B_1} x_2^+ \chi_{\{x:\pi/6 < \theta < 5\pi/6\}} \, dx$, *then the free boundary is in a neighborhood of* x^0 *the union of two* C^1-*graphs of functions* $\eta_1 : (x_1^0 - \delta, x_1^0] \to$ **R** *and* $\eta_2 : [x_1^0, x_1^0 + \delta) \to$ **R** *which are both continuously differentiable up to* x_1^0 *and satisfy* $\eta_1'(x_1^0) = -1/\sqrt{3}$ *and* $\eta_2'(x_1^0) = 1/\sqrt{3}$.

(ii) *Else* $M_{x^0,u}(0+) = 0$, *and* $\sigma_1(t) \neq x_1^0$ *in* $(-t_1, t_1) \setminus \{0\}$, *and* $\sigma_1 - x_1^0$ *does not change its sign at* $t = 0$, *and*

$$\lim_{t \to 0} \frac{\sigma_2(t)}{\sigma_1(t) - x_1^0} = 0.$$

If we assume in addition that either $\{u > 0\}$ *is a supergraph of a function in the* x_2-*direction or that* $\{u > 0\}$ *is a Lipschitz set, then the set* S^u *of stagnation points is locally in* Ω *a finite set, and at each stagnation point* x^0 *the statement in* (i) *holds.*

Proof We first show that the set Σ^u is empty. Suppose towards a contradiction that there exists $x^0 \in \Sigma^u$. From Theorem 9 we infer that there exists an integer $N(x^0) \geq 2$ such that

$$v_r(x) := \frac{u(x^0 + rx)}{\sqrt{r^{-1} \int_{\partial B_r(x^0)} u^2 \, d\mathcal{H}^1}} \tag{4.58}$$

$$\to \frac{\rho^{N(x^0)} |\sin(N(x^0) \min(\max(\theta, 0), \pi))|}{\sqrt{\int_0^\pi \sin^2(N(x^0)\theta) d\theta}} \quad \text{as } r \to 0+,$$

strongly in $W_{\text{loc}}^{1,2}(B_1 \setminus \{0\})$ and weakly in $W^{1,2}(B_1)$, where $x = (\rho \cos\theta, \rho \sin\theta)$. On the other hand, Theorem 4(ii) implies that for any ball $\tilde{B} \subset\subset B_1 \cap \{x_2 > 0\}$, $v_r > 0$ in $\tilde{B}$. Consequently (see (4.53))

$$|\Delta v_r| \leq C_1 r^2 v_r \text{ in } \tilde{B}$$

for sufficiently small r. It follows that v_0 is harmonic in $\tilde{B}$, contradicting (4.58) in view of $N(x^0) \geq 2$. Hence Σ^u is indeed empty.

Let us consider the case $M_{x^0,u}(0+) = \int_{B_1} x_2^+ \chi_{\{x:\pi/6<\theta<5\pi/6\}} dx$. From Theorem 4 and from the isolation property [38, Proposition 5.4] we infer that

$$\frac{u(x^0 + rx)}{r^{3/2}} \to \frac{\sqrt{2}}{3} \rho^{3/2} \cos(\frac{3}{2}(\min(\max(\theta, \frac{\pi}{6}), \frac{5\pi}{6}) - \frac{\pi}{2})) \quad \text{as } r \to 0+,$$

$$(4.59)$$

strongly in $W_{loc}^{1,2}(\mathbf{R}^2)$ and locally uniformly on $\mathbf{R}^2$, where $x = (\rho \cos \theta, \rho \sin \theta)$.

We assume for simplicity that $x^0 = 0$. We will show that in a neighborhood of 0 the free boundary is the union of two C^1-graphs $\eta_1 : (-\delta, 0] \to \mathbf{R}$ and $\eta_2 : [0, \delta) \to \mathbf{R}$ which are both continuously differentiable up to 0 and satisfy $\eta_1'(0) = -1/\sqrt{3}$ and $\eta_2'(0) = 1/\sqrt{3}$: as the proofs for $x_1 > 0$ and $x_1 < 0$ are similar, we will give only the proof for $x_1 > 0$.

For

$$v(x) := \frac{u(\rho x)}{\rho^{3/2}}$$

we have that

$$\Delta v(x) = -\sqrt{\rho} f(u(\rho x)) \text{ for } v(x) > 0,$$

$$|\nabla v(x)|^2 = x_2 \text{ for } x \in \partial\{v > 0\}.$$

Scaling once more for $\xi \in \partial B_1 \cap \partial\{v > 0\}$, which implies that for ρ small enough, $\xi_2 \geq \frac{1}{10}$, we obtain for

$$w(x) := \frac{v(\xi + rx)}{\xi_2 r}$$

that

$$\Delta w(x) = -\frac{\sqrt{\rho} r}{\xi_2} f(\rho \xi + r\rho x) \text{ for } w(x) > 0,$$

$$|\nabla w(x)|^2 = 1 + \frac{r x_2}{\xi_2} \text{ for } x \in \partial\{w > 0\}.$$

We are going to use a flatness-implies-regularity result of [13]. Note that although not stated in [13], [13, Lemma 4.1] yields as in the proof of [13, Theorem 1.1] that for each $\epsilon \in (0, \epsilon_0)$

$$\max(x \cdot \bar{\nu} - \epsilon, 0) \leq w \leq \max(x \cdot \bar{\nu} + \epsilon, 0) \text{ in } B_1 \qquad (4.60)$$

Fig. 4.13 Do examples like this exist?

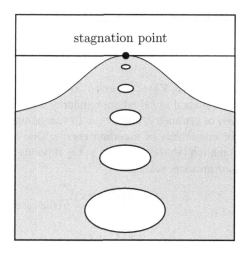

stagnation point

implies that the outward unit normal ν^w on the free boundary $\partial\{w > 0\}$ satisfies

$$|\nu^w(0) - \bar{\nu}| \leq C\epsilon^2.$$

Note that $\nu^w(0) = \nu(\rho\xi)$. Since (4.60) is by (4.59) satisfied for $\bar{\nu} = (1/2, -\sqrt{3}/2)$, $r = r(\epsilon)$ and every sufficiently small $\rho > 0$, we obtain that the outward unit normal $\nu(x)$ on $\partial\{u > 0\}$ converges to $\bar{\nu}$ as $x \to 0$, $x_1 > 0$. It follows that the present curve component is the graph of a C^1-function (up to $x_1 = x_1^0$) in the x_2-direction.

The remaining statements of the Theorem follow from Theorem 4.

The reader may wonder whether the injective curve assumption is necessary. This is an open problem, however it does not seem easy to rule out a scenario like in the following picture showing possible level sets of ψ (Fig. 4.13).

4.14 Extensions

1. In [39] we extended our methods to cover the axisymmetric three-dimensional case: Consider the steady axisymmetric Euler equations for a fluid (incompressible and with zero vorticity) with a free surface acted on only by gravity. Using cylindrical coordinates and the Stokes stream function ψ (see for example [16, Exercise 4.18 (ii)]), we obtain the free boundary problem

$$\text{div}\left(\frac{1}{x_1}\nabla\psi(x_1, x_2)\right) = 0 \text{ in the water phase } \{\psi > 0\} \tag{4.61}$$

$$\frac{1}{x_1^2}|\nabla\psi(x_1, x_2)|^2 = -x_2 \text{ on the free surface } \partial\{\psi > 0\};$$

here the original velocity field

$$V(X, Y, Z) = \left(-\frac{1}{x_1} \partial_2 \psi \cos \vartheta, \ -\frac{1}{x_1} \partial_2 \psi \sin \vartheta, \ \frac{1}{x_1} \partial_1 \psi \right),$$

where $(X, Y, Z) = (x_1 \cos \vartheta, x_1 \sin \vartheta, x_2)$. The free boundary problem (4.61) has been studied in [3] where regularity away from the degenerate sets $\{x_1 = 0\}$ (the axis of symmetry) and $\{x_2 = 0\}$ (containing all stagnation points) has been shown for minimizers of a certain energy. Due to the degeneracy of the free boundary condition $|\nabla \psi(x_1, x_2)|^2 = x_1^2 x_2$ at points $x^0 = (x_1^0, x_2^0)$ with $x_1^0 x_2^0 = 0$, we obtain *four* invariant scalings

$$\frac{\psi(x^0 + rx)}{r} \quad \text{in the case } x_1^0 \neq 0 \text{ and } x_2^0 \neq 0,$$

$$\frac{\psi(x^0 + rx)}{r^{\frac{3}{2}}} \quad \text{in the case } x_1^0 \neq 0 \text{ and } x_2^0 = 0,$$

$$\frac{\psi(x^0 + rx)}{r^2} \quad \text{in the case } x_1^0 = 0 \text{ and } x_2^0 \neq 0,$$

$$\frac{\psi(x^0 + rx)}{r^{\frac{5}{2}}} \quad \text{in the case } x_1^0 = x_2^0 = 0.$$

Note that the velocity (in the moving frame) would scale like $1, |x|^{\frac{1}{2}}, 1, |x|^{\frac{1}{2}}$ in the respective cases. In [39] we determined the profile of the scaled solution as $r \to 0$ (Proposition 3): In the case $x_1^0 \neq 0$ and $x_2^0 \neq 0$ the only asymptotics possible is constant velocity flow parallel to the free surface. In the case $x_1^0 \neq 0$ and $x_2^0 = 0$ the only asymptotics possible is the well-known Stokes corner flow (see [5, 29, 37]). Due to the perturbed equation the situation is actually not unlike the two-dimensional problem in the presence of vorticity (see [11, 12, 38] for two-dimensional results in the presence of vorticity). In the case $x_1^0 = 0$ and $x_2^0 \neq 0$ the only asymptotics possible is constant velocity flow in the gravity direction. This suggests the *possibility of air cusps* pointing in the gravity direction (Fig. 4.14).

In the case $x_1^0 = x_2^0 = 0$ the only asymptotics possible is the *Garabedian pointed bubble solution* with water above air (cf. [17], Fig. 4.15). This comes at first as a surprise as it means that there is no nontrivial asymptotic profile at all with air above water and with the invariant scaling.

Fig. 4.14 Air cusp pointing
in the gravity direction

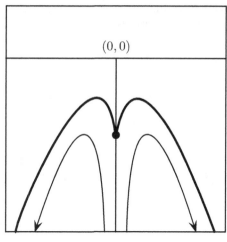

$(0,0)$

Fig. 4.15 Garabedian
pointed bubble

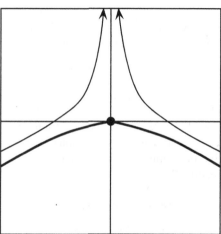

Assuming that the surface is given by an injective curve and assuming also a strict
Bernstein inequality (corresponding to a Rayleigh-Taylor condition) we obtain the
following result:

In the case $x_1^0 \neq 0$ and $x_2^0 = 0$ the only asymptotics possible are the well-known
Stokes corner (an angle of opening $120°$ in the direction of the axis of symmetry),
and a horizontal point.

In the case $x_1^0 = 0$ and $x_2^0 < 0$ the only asymptotics possible are cusps in the
direction of the axis of symmetry.

In the case $x_1^0 = x_2^0 = 0$ the only asymptotics possible are the *Garabedian
pointed bubble asymptotics* (an angle of opening $\approx 114.799°$ with water above air),
and a *horizontal point*.

A fine analysis of the velocity profile in the last case ($x_1^0 = x_2^0 = 0$ and
a horizontal point) is no mean feat, and we confine ourselves to the case of

Fig. 4.16 Dynamics
suggested by our analysis

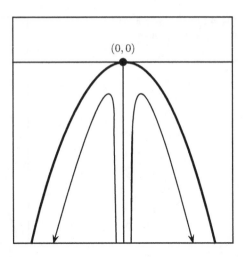

air above water. Here we prove (Theorem 9) that the velocity scales almost like $\sqrt{X^2 + Y^2 + Z^2}$ and is asymptotically given by the velocity field

$$V(\sqrt{X^2 + Y^2}, Z) = c(-\sqrt{X^2 + Y^2}, 2Z),$$

where c is a nonzero constant (Fig. 4.16).

2. In [33] we considered a stationary flow of an incompressible, inviscid perfectly conducting liquid coupled to an electric field. In our setting the Electro-HydroDynamic equations (EHD equations) simplify (see [18, (10)] as well as [46, Section 2]) to

$$\Delta\phi = 0 \text{ in the gas,} \tag{4.62}$$

$$\Delta V = 0 \text{ in the fluid,} \tag{4.63}$$

$$|\nabla\phi|^2 - |\nabla V|^2 = \kappa + B \text{ on the free surface of the fluid,} \tag{4.64}$$

$$\phi = 0 \text{ on the free surface of the fluid,} \tag{4.65}$$

$$\nabla V \cdot \nu = 0 \text{ on the free surface of the fluid,} \tag{4.66}$$

where B is a constant, V is the velocity potential of the stationary fluid, ν is the outward pointing unit normal and κ the mean curvature on the boundary of the fluid phase. Note that we choose the sign of the mean curvature of the boundary of the set A so that κ is positive on convex portions of ∂A. Viewed as a free boundary problem, problem (4.62)–(4.66) was new, so there were no results from that perspective. Possible reasons for the lack of results may be the "bad" sign of the mean curvature (explained in more detail below) as well as the Neumann boundary condition (4.66). While there are many free boundary results for level sets, there are relatively few on problems without this property.

Still there were related free boundary/free discontinuity problems: In [9], the authors study the free boundary problem

$$\Delta u = 0 \text{ in } \Omega \cap (\{u > 0\} \cup \{u < 0\}), \tag{4.67}$$

$$|\nabla u^+|^2 - |\nabla u^-|^2 = -\kappa \text{ on the free surface } \partial\{u \le 0\} \cap \Omega. \tag{4.68}$$

Even in the case $u^- \equiv 0$, problem (4.67)–(4.68) differs from (4.62)–(4.66) by the sign of the mean curvature. This becomes clearer when comparing the energies of the two problems: In case of one-phase solutions ($u^- \equiv 0$) of (4.67)–(4.68), the energy takes the form

$$P_\Omega(\{u > 0\}) + \int_{\Omega \cap \{u > 0\}} |\nabla u|^2,$$

where $P_\Omega(\{u > 0\})$ is the perimeter of the set $\{u > 0\}$ relative to the fixed domain of definition Ω. In case of one-phase solutions ($V \equiv 0$) of problem (4.62)–(4.66), where we extend ϕ by the value 0 to the fluid phase and we consider $B = 0$, the energy takes the form

$$- P_\Omega(\{\phi > 0\}) + \int_{\Omega \cap \{\phi > 0\}} |\nabla \phi|^2. \tag{4.69}$$

As a consequence, constructing solutions by minimizing the energy makes perfect sense for (4.67)–(4.68) and leads in dimension $n \le 7$ to regular solutions (see [22]), while there obviously exist no minimizers of energy (4.69). Moreover, critical points of the energy (4.69) may have singularities even in dimension 2. An example of a singularity would be the real part of the complex root, multiplied by a suitable positive constant. It is not difficult to find more evidence underlining the drastic difference in the qualitative behaviour of solutions of (4.67)–(4.68) and (4.62)–(4.66).

In two dimensions system (4.62)–(4.66) is under certain assumptions equivalent to critical points of the Mumford-Shah functional (concerning the Mumford-Shah functional see for example [4] and [24]): the Mumford-Shah equations are (up to terms of lower order)

$$\Delta m = 0 \text{ in } \Omega \setminus S_m,$$

$$[|\nabla m|^2] = \kappa \text{ on the free discontinuity set } \Omega \cap S_m,$$

$$\nabla m \cdot \nu = 0 \text{ on the free discontinuity set } \Omega \cap S_m,$$

where $[|\nabla m|^2]$ denotes the jump of $|\nabla m|^2$. The sign of the jump and that of the mean curvature are chosen such that $m = 0$ in a component D of Ω implies that

$$|\nabla m|^2 = -\kappa \text{ on } \Omega \cap \partial(\Omega \setminus D).$$

Note that the homogeneous Dirichlet condition (4.66) is replaced by a homogeneous Neumann condition in the Mumford-Shah problem. That means that the Mumford-Shah problem can be applied either to one-phase solutions v of (4.62)–(4.66), in which case $\phi \equiv 0$. Another possibility in two dimensions is to take under certain assumptions the harmonic conjugate of ϕ which would then, combined with V, yield a solution of the Mumford-Shah equations. This also means that in two dimensions the paper [43] is directly applicable to (4.62)–(4.66).

In [33] we proved for a variational solution u the following: Suppose that $0 \in \partial\{u > 0\}$ and that $\partial\{u > 0\} \cap B_1^+$ is, in a neighborhood of 0, a continuous injective curve $\sigma : I \to \mathbf{R}^2$, where I is an interval of $\mathbf{R}$ containing 0 such that $\sigma = (\sigma_1, \sigma_2)$ and $\sigma(0) = 0$. Suppose, additionally, that there exist $r_0, C > 0$ such that for $r \in (0, r_0)$,

$$Cr^{2\gamma-2} \le \int_{B_1^+} x_1 |\nabla u^+(rx)|^2. \tag{4.70}$$

(Or, analogously,

$$Cr^{2\gamma} \le \int_{B_1^+} \frac{1}{x_1} |\nabla u^-(rx)|^2)$$

Then,

$$\lim_{t \to 0+} \left| \frac{\sigma_1(t)}{\sigma_2(t)} \right| = 0,$$

that is, the free boundary is asymptotically cusp-shaped (Fig. 4.17).

Fig. 4.17 The fluid is asymptotically cusp-shaped

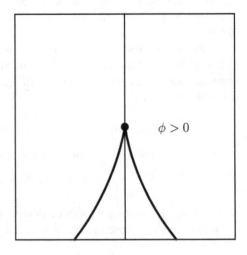

$\phi > 0$

References

1. F.J. Jr Almgren, *Almgren's big regularity paper*, volume 1 of *World Scientific Monograph Series in Mathematics* (World Scientific Publishing Co. Inc., River Edge, 2000). Q-valued functions minimizing Dirichlet's integral and the regularity of area-minimizing rectifiable currents up to codimension 2, With a preface by J.E. Taylor, V. Scheffer
2. H.W. Alt, L.A. Caffarelli, Existence and regularity for a minimum problem with free boundary. J. Reine Angew. Math. **325**, 105–144 (1981)
3. H.W. Alt, L.A. Caffarelli, A. Friedman, Jet flows with gravity. J. Reine Angew. Math. **331**, 58–103 (1982)
4. L. Ambrosio, N. Fusco, D. Pallara. *Functions of Bounded Variation and Free Discontinuity Problems*. Oxford Mathematical Monographs (The Clarendon Press Oxford University Press, New York, 2000)
5. C.J. Amick, L.E. Fraenkel, J.F. Toland, On the Stokes conjecture for the wave of extreme form. Acta Math. **148**, 193–214 (1982)
6. C.J. Amick, J.F. Toland, On solitary water-waves of finite amplitude. Arch. Ration. Mech. Anal. **76**, 9–95 (1981)
7. J. Andersson, G.S. Weiss, A parabolic free boundary problem with Bernoulli type condition on the free boundary. J. Reine Angew. Math. **627**, 213–235 (2009)
8. J. Andersson, H. Shahgholian, G.S. Weiss, A variational linearization technique in free boundary problems applied to a two-phase bernoulli problem. Preprint (2017)
9. I. Athanasopoulos, L.A. Caffarelli, C. Kenig, S. Salsa, An area-Dirichlet integral minimization problem. Commun. Pure Appl. Math. **54**(4), 479–499 (2001)
10. L.A. Caffarelli, D. Jerison, C.E. Kenig, Global energy minimizers for free boundary problems and full regularity in three dimensions, in *Noncompact problems at the intersection of geometry, analysis, and topology*. Contemp. Math. **350** (American Mathematical Society, Providence, pp. 83–97, 2004)
11. A. Constantin, W. Strauss, Exact steady periodic water waves with vorticity. Commun. Pure Appl. Math. **57**(4), 481–527 (2004)
12. A. Constantin, W. Strauss, Periodic traveling gravity water waves with discontinuous vorticity. Arch. Ration. Mech. Anal. **202**(1), 133–175 (2011)
13. D. de Silva, Free boundary regularity for a problem with right hand side. Interfaces Free Bound. **13**, 223–238 (2011)
14. L.C. Evans, S. Müller, Hardy spaces and the two-dimensional Euler equations with nonnegative vorticity. J. Am. Math. Soc. **7**(1), 199–219 (1994)
15. L.C. Evans, *Weak convergence methods for nonlinear partial differential equations*, volume 74 of *CBMS Regional Conference Series in Mathematics* (Published for the Conference Board of the Mathematical Sciences, Washington, DC, 1990)
16. L.E. Fraenkel, *An Introduction to Maximum Principles and Symmetry in Elliptic Problems*, volume 128 of *Cambridge Tracts in Mathematics* (Cambridge University Press, Cambridge, 2000)
17. P.R. Garabedian, A remark about pointed bubbles. Commun. Pure Appl. Math. **38**(5), 609–612 (1985)
18. S. Grandison, J.-M. Vanden-Broeck, D.T. Papageorgiou, T. Miloh, B. Spivak, Axisymmetric waves in electrohydrodynamic flows. J. Engrg. Math. **62**(2), 133–148 (2008)
19. N. Garofalo, F.-H. Lin, Monotonicity properties of variational integrals, A_p weights and unique continuation. Indiana Univ. Math. J. **35**(2), 245–268 (1986)
20. D. Gilbarg, N.S. Trudinger, *Elliptic Partial Differential Equations of Second Order*, volume 224 of *Grundlehren der Mathematischen Wissenschaften [Fundamental Principles of Mathematical Sciences]* 2nd edn. (Springer, Berlin, 1983)
21. E. Giusti, *Minimal Surfaces and Functions of Bounded Variation*, volume 80 of *Monographs in Mathematics* (Birkhäuser Verlag, Basel, 1984)

22. H. Jiang, C.J. Larsen, L. Silvestre, Full regularity of a free boundary problem with two phases. Calc. Var. Partial Differ. Equ. **42**(3–4), 301–321 (2011)

23. G. Keady, J. Norbury, On the existence theory for irrotational water waves. Math. Proc. Camb. Philos. Soc. **83**, 137–157 (1978)

24. H. Koch, G. Leoni, M. Morini, On optimal regularity of free boundary problems and a conjecture of De Giorgi. Commun. Pure Appl. Math. **58**(8), 1051–1076 (2005)

25. J.P. Krasovskiĭ, On the theory of steady-state waves of finite amplitude. Ž. Vyčisl. Mat. i Mat. Fiz. **1**, 836–855 (1961)

26. J.B. McLeod, The Stokes and Krasovskii conjectures for the wave of greatest height. Stud. Appl. Math. **98**, 311–333 (1997)

27. S. Luckhaus, Solutions for the two-phase Stefan problem with the Gibbs-Thomson law for the melting temperature. Eur. J. Appl. Math. **1**(2), 101–111 (1990)

28. F. Pacard, Partial regularity for weak solutions of a nonlinear elliptic equation. Manuscripta Math. **79**(2), 161–172 (1993)

29. P.I. Plotnikov, Proof of the Stokes conjecture in the theory of surface waves. Stud. Appl. Math. **108**(2), 217–244 (2002) Translated from Dinamika Sploshn. Sredy No. 57 (1982), 41–76 [MR0752600 (85f:76036)]

30. P.I. Plotnikov, J.F. Toland, Convexity of Stokes waves of extreme form. Arch. Ration. Mech. Anal. **171**, 349–416 (2004)

31. P. Price, A monotonicity formula for Yang-Mills fields. Manuscripta Math. **43**(2–3), 131–166 (1983)

32. R.M. Schoen, Analytic aspects of the harmonic map problem, in *Seminar on Nonlinear Partial Differential Equations (Berkeley, Calif., 1983)*, volume 2 of *Mathematical Sciences Research Institute Publications* (Springer, New York, 1984), pp. 321–358

33. M. Smit Vega Garcia, E. Vărvărucă, G.S. Weiss, Singularities in axisymmetric free boundaries for electrohydrodynamic equations. Arch. Ration. Mech. Anal. **222**(2), 573–601 (2016)

34. G.G. Stokes, *Considerations Greatest Height of Oscillatory Irrotational Waves Which can be Propagated Without Change of Form*, volume 1 of *Math. and Phys. Papers* (Cambridge University Press, Cambridge, 1880)

35. J.F. Toland, On the existence of a wave of greatest height and Stokes's conjecture. Proc. R. Soc. Lond. Ser. A **363**, 469–485 (1978)

36. E. Varvaruca, On the existence of extreme waves and the Stokes conjecture with vorticity. J. Differ. Equ. **246**(10), 4043–4076 (2009)

37. E. Varvaruca, G.S. Weiss, A geometric approach to generalized Stokes conjectures. Acta Math. **206**, 363–403 (2011)

38. E. Varvaruca, G.S. Weiss, The Stokes conjecture for waves with vorticity. Ann Inst H Poincaré Anal Non Linéaire **29**(6), 861–885 (2012)

39. E. Varvaruca, G.S. Weiss, Singularities of steady axisymmetric free surface flows with gravity. Commun. Pure Appl. Math. **67**(8), 1263–1306 (2014)

40. G.S. Weiss, Partial regularity for weak solutions of an elliptic free boundary problem. Commun. Partial Differ. Equ. **23**(3–4), 439–455 (1998)

41. G.S. Weiss, Partial regularity for a minimum problem with free boundary. J. Geom. Anal. **9**(2), 317–326 (1999)

42. G.S. Weiss, A singular limit arising in combustion theory: fine properties of the free boundary. Calc. Var. Partial Differ. Equ. **17**(3), 311–340 (2003)

43. G. Weiss, G. Zhang, A free boundary approach to two-dimensional steady capillary gravity water waves. Arch. Ration. Mech. Anal. **203**, 747–768 (2012)

44. T.H. Wolff, Plane harmonic measures live on sets of σ-finite length. Ark Mat **31**(1), 137–172 (1993)

45. T.H. Wolff, Counterexamples with harmonic gradients in $\mathbf{R}^3$, in *Essays on Fourier analysis in honor of Elias M. Stein* (Princeton, 1991), Princeton Mathematical Series, vol. 42 (Princeton University Press, Princeton, 1995), pp. 321–384

46. N.M. Zubarev, Criteria for hard excitation of electrohydrodynamic instability of the free surface of a conducting fluid. Physica D **152–153**, 787–793 (2001)

LECTURE NOTES IN MATHEMATICS   🐎 Springer

Editors in Chief: J.-M. Morel, B. Teissier;

Editorial Policy

1. Lecture Notes aim to report new developments in all areas of mathematics and their applications – quickly, informally and at a high level. Mathematical texts analysing new developments in modelling and numerical simulation are welcome.

   Manuscripts should be reasonably self-contained and rounded off. Thus they may, and often will, present not only results of the author but also related work by other people. They may be based on specialised lecture courses. Furthermore, the manuscripts should provide sufficient motivation, examples and applications. This clearly distinguishes Lecture Notes from journal articles or technical reports which normally are very concise. Articles intended for a journal but too long to be accepted by most journals, usually do not have this "lecture notes" character. For similar reasons it is unusual for doctoral theses to be accepted for the Lecture Notes series, though habilitation theses may be appropriate.

2. Besides monographs, multi-author manuscripts resulting from SUMMER SCHOOLS or similar INTENSIVE COURSES are welcome, provided their objective was held to present an active mathematical topic to an audience at the beginning or intermediate graduate level (a list of participants should be provided).

   The resulting manuscript should not be just a collection of course notes, but should require advance planning and coordination among the main lecturers. The subject matter should dictate the structure of the book. This structure should be motivated and explained in a scientific introduction, and the notation, references, index and formulation of results should be, if possible, unified by the editors. Each contribution should have an abstract and an introduction referring to the other contributions. In other words, more preparatory work must go into a multi-authored volume than simply assembling a disparate collection of papers, communicated at the event.

3. Manuscripts should be submitted either online at www.editorialmanager.com/lnm to Springer's mathematics editorial in Heidelberg, or electronically to one of the series editors. Authors should be aware that incomplete or insufficiently close-to-final manuscripts almost always result in longer refereeing times and nevertheless unclear referees' recommendations, making further refereeing of a final draft necessary. The strict minimum amount of material that will be considered should include a detailed outline describing the planned contents of each chapter, a bibliography and several sample chapters. Parallel submission of a manuscript to another publisher while under consideration for LNM is not acceptable and can lead to rejection.

4. In general, **monographs** will be sent out to at least 2 external referees for evaluation.

   A final decision to publish can be made only on the basis of the complete manuscript, however a refereeing process leading to a preliminary decision can be based on a pre-final or incomplete manuscript.

   Volume Editors of **multi-author works** are expected to arrange for the refereeing, to the usual scientific standards, of the individual contributions. If the resulting reports can be forwarded to the LNM Editorial Board, this is very helpful. If no reports are forwarded or if other questions remain unclear in respect of homogeneity etc, the series editors may wish to consult external referees for an overall evaluation of the volume.

5. Manuscripts should in general be submitted in English. Final manuscripts should contain at least 100 pages of mathematical text and should always include

   – a table of contents;
   – an informative introduction, with adequate motivation and perhaps some historical remarks: it should be accessible to a reader not intimately familiar with the topic treated;
   – a subject index: as a rule this is genuinely helpful for the reader.
   – For evaluation purposes, manuscripts should be submitted as pdf files.

6. Careful preparation of the manuscripts will help keep production time short besides ensuring satisfactory appearance of the finished book in print and online. After acceptance of the manuscript authors will be asked to prepare the final LaTeX source files (see LaTeX templates online: https://www.springer.com/gb/authors-editors/book-authors-editors/manuscriptpreparation/5636) plus the corresponding pdf- or zipped ps-file. The LaTeX source files are essential for producing the full-text online version of the book, see http://link.springer.com/bookseries/304 for the existing online volumes of LNM). The technical production of a Lecture Notes volume takes approximately 12 weeks. Additional instructions, if necessary, are available on request from lnm@springer.com.

7. Authors receive a total of 30 free copies of their volume and free access to their book on SpringerLink, but no royalties. They are entitled to a discount of 33.3 % on the price of Springer books purchased for their personal use, if ordering directly from Springer.

8. Commitment to publish is made by a *Publishing Agreement*; contributing authors of multiauthor books are requested to sign a *Consent to Publish form*. Springer-Verlag registers the copyright for each volume. Authors are free to reuse material contained in their LNM volumes in later publications: a brief written (or e-mail) request for formal permission is sufficient.

Addresses:
Professor Jean-Michel Morel, CMLA, École Normale Supérieure de Cachan, France
E-mail: moreljeanmichel@gmail.com

Professor Bernard Teissier, Équipe Géométrie et Dynamique,
Institut de Mathématiques de Jussieu – Paris Rive Gauche, Paris, France
E-mail: bernard.teissier@imj-prg.fr

Springer: Ute McCrory, Mathematics, Heidelberg, Germany,
E-mail: lnm@springer.com

Printed in the United States
by Baker & Taylor Publisher Services